COMPUTER APPLICATIONS

IN AGRICULTURE

AND AGRIBUSINESS

MICHAEL E. NEWMAN
Assistant Professor
Department of Agricultural Education
and Experimental Statistics
Mississippi State University

Interstate Publishers, Inc.
Danville, Illinois

COMPUTER APPLICATIONS IN AGRICULTURE AND AGRIBUSINESS. COPYRIGHT© 1994 BY INTERSTATE PUBLISHERS, INC. All rights reserved.

Order from
Interstate Publishers, Inc.
510 N. Vermilion St.
P. O. Box 50
Danville, IL 61834-0050
Phone : (800) 843-4774
FAX: (217) 446-9706

Library of Congress Catalog Card No. 93-80108

ISBN 0-8134-2976-5

1 2 3
4 5 6
7 8 9

Foreword

Computer Applications for Agriculture and Agribusiness contains seven chapters from the book, **Agribusiness Management and Entrepreneurship** (1994), also offered by Interstate Publishers, Inc.

These chapters were produced as a separate text due to teacher requests. After initial reviews, several teachers and agricultural education curriculum experts recommended these chapters be published for use as a supplemental text for other agricultural education courses.

This book was written to provide an introduction to computers for the novice. It is appropriate for students at the secondary, post-secondary, or adult level. Students in all areas of agriculture—agriscience, production, horticulture, forestry, aquaculture, products processing, and many others—find the computer useful as a management tool. The concepts and applications presented in this text can be used in any of these areas.

The first chapter provides a discussion of the history of computers and the principles of computer operation. The next chapter provides an overview of operating systems and environments. The last five chapters present detailed descriptions of the five most commonly used computer applications: word processors, electronic spreadsheets, database managers, graphics, and networking/communications.

This text provides useful, general information about computers and computer applications. To get the most benefit from the book, the reader should put the information to use with a computer and software. The documentation will provide more specific instruction about the software.

Michael E. Newman

Contents

Chapter		Page
	Preface	iii
1	Principles of Computer Operations	1
2	Disk Operating Systems	47
3	Word Processing	81
4	Electronic Spreadsheets/Recordkeeping	103
5	Database Management	121
6	Graphics	133
7	Communications and Networking	151
	Glossary	167
	Index	172

Chapter 1

PRINCIPLES OF COMPUTER OPERATION

Computers are everywhere! Most people cannot spend a day without coming in contact with some type of computer. A trip to the grocery store includes having the items scanned by a computer component. Modern cash registers have a computer inside. A catalog order will be punched into a computer. Bills that come in the mail almost surely have been generated by a computer. If a credit card is used for a gasoline purchase, the cashier sends the information via telephone to a computer. On some dairy farms, a computer reads the tag on an individual cow, determines how much feed she has eaten that day and how much more she is allowed to eat.

Many individuals use computers for word processing, keeping records, or generating financial statements to help them accomplish their work on a given day. Since 1981, computers have become affordable for almost every business and many individuals. These people are taking advantage of the benefits of using a computer to make them more efficient.

In the previous chapters of this book, many agribusiness management functions and strategies have been discussed. The computer can help the agribusiness manager with almost all of these functions. This chapter provides an introduction to computers and how they work for individuals who have little background in using computers.

OBJECTIVES

1. Define computers.
2. Discuss the history of computing.
3. Describe how computers work.
4. Describe the parts of a computer (hardware).

5. Identify the various classifications of software.
6. Discuss the factors to consider when buying a computer.

COMPUTERS AS A BUSINESS TOOL

A computer, simply defined, is an electronic device that executes programs. A *program* is a detailed set of instructions written by a computer *programmer* which may be simple or complex.

The computer is useful as a tool in business and other areas because the programs it executes can perform a wide variety of tasks. If the pro-

TERMS

artificial intelligence
application
backup
binary logic
bit
bitmapping
byte
commercial software
cathode ray tube (CRT)
cursor
cursor control keys
daisy wheel printer
data
data files
disk drive
documentation
dot matrix printer
dot pitch
floppy disk
FORTRAN
freeware
function keys

gigabyte (GB)
graphical user interface (GUI)
graphics
hard disk
hardware
icons
inkjet printer
input device
integrated software
kilobyte (K)
laser printer
liquid crystal display (LCD)
machine language
math coprocessor
megabyte (MB or Meg)
memory
microchip
microprocessor
millions of instructions per second (MIPS)
mouse

near letter quality
numeric keypad
operating system
output device
peripherals
pixel
program
programmer
random access memory (RAM)
resolution
read-only memory (ROM)
semi-conductor
software
software piracy
terminal
terminate-and-stay-resident (TSR)
transistor
user-supported software
virus

PRINCIPLES OF COMPUTER OPERATION   3

Figure 17-1. In the 1990s, most executives have computers available for ready access.

grammer can imagine it and write the program, the computer can perform the task (with a few limitations).

A person can do anything a computer can do, but there are several ways in which a person or business can use computers to become more efficient. First, a computer, when programmed correctly, never gives a wrong answer. Second, unlike humans, the computer doesn't get fatigued, even when doing mundane tasks. A computer will perform a simple arithmetic problem thousands of times without complaining or getting tired. Third, the computer does these things at incredible speeds. A computer can add two numbers in $1/2{,}000{,}000$ of a second—the time it takes light to travel 12 feet.

Computers are not smart, but they do follow directions. A computer is only as smart as the people who programmed it and the people who operate it. Anyone who has had problems with service from a business has probably heard something like, "I'm sorry, our computer is down," or "Our computer won't do that for us." These people are allowing computers to dictate their work *to them* instead of doing work *for them*. Used properly, the computer is a tool that makes work easier, not harder.

HISTORY OF COMPUTING

The earliest computing involved counting and developing a place system, probably to keep track of time. Before long, humans felt a need for assistance in their computations. Many devices were created to aid in computing long before the computer, as we know it today, came along. These early machines and computing devices, however, did lead us to the modern day computer.

EARLY COMPUTING DEVICES

The abacus, developed around 1000 B.C., was probably the first computing device, and almost certainly was the first portable computing device. It used beads strung to a frame and worked on the principle of place value, with different beads having different values. The beads were called calculi. Calculi is the plural of calculus, which is the area of math that deals with small differences. Calculus is the root word for calculate and calculator.

Figure 17-2. An abacus—probably the earliest portable computing device.

Although the abacus had been used in the Orient for many years, it wasn't until the early 17th century that Europeans invented any type of calculating machine. They caught up quickly, however. In 1617, Napier, a Scottish nobleman, developed an instrument for performing multiplication and division. The instrument matched up numbers on rods that were side by side, then the results were read from the other side. This device, called Napier's Bones, was the prototype of the slide rule.

Around 1642, Blaise Pascal, a French scientist and philosopher, invented the first gear-driven or mechanical calculator. He was 19 at the time and built the machine to help his father, who was a mathematician. The machine was called a Pascaline. Each gear had ten teeth representing a number between 1 and 10. When one gear completed a rotation, the next gear would move one notch, much like a car odometer. This machine did addition and subtraction. This was the first machine to raise a stir because it caused fear of automation and thus unemployment.

Leibniz, a philosopher and mathematician who invented differential and integral calculus (independent of Sir Isaac Newton), invented a device called Leibniz's Wheel in 1673. It was a gear-driven machine that was superior to Pascal's because it could also multiply and divide. This was basically the forerunner of the hand-operated adding machines and calculators used into the 1970s.

Another device important to computing was Jacquard's Loom, developed around 1801. It had a device to control the weaving pattern of a loom. Jacquard used a perforated card to control the selection of threads used to create a pattern of thread. Changing cards changed the pattern. This invention caused widespread riots because of fear of automation. Workers threw boots (*sabots*) into looms. This is where the word sabotage originated. Card-driven equipment is still used in the textile industry today.

For over 40 years, from around 1790 until into the 1830s, Charles Babbage and Ada Byron worked on the Analytical and Difference Engines. Babbage is considered the Father of Computers because of his invention of these machines, while Byron is considered the first computer programmer because of her analyses and explanations of his work. The difference engine, a model of which was built, solved polynomial equations by the method of differences. The analytical engine, never built, was designed to be a general purpose computing device.

The analytical engine design was the prototype of the modern computer. It contained five parts common in today's computers: (1) an *input device* using punched cards, an idea taken from Jaquard's Loom; (2) a processor, which was a mill containing hundreds of vertical axles and thousands of gears, that stood over ten feet tall; (3) a control unit, in this case a barrel-like device with slats and studs, operating like a complex player piano; (4) a storage device with more axles and gears, designed to hold 100 40-digit numbers; and (5) an *output device* that made a set of plates designed to fit a printing press. Babbage was so far ahead of his time that the machining and tooling available was not good enough to build his machine.

Necessity was the mother of invention in the case of the Hollerith Card, invented around 1890. Herman Hollerith, an employee of the U.S. Census Bureau, won a competition to develop a method that would be faster than hand-tabulation for use in the 1890 Census. He expanded upon Jacquard's idea and developed the punched card, a code, and a machine to read the code. With the punched card, the 1890 Census results were announced within six weeks (whereas the 1880 Census took over seven years). Hollerith left the Census Bureau and formed his own company, the Tabulating Machine Company, known today as International Business Machines or IBM. Cards

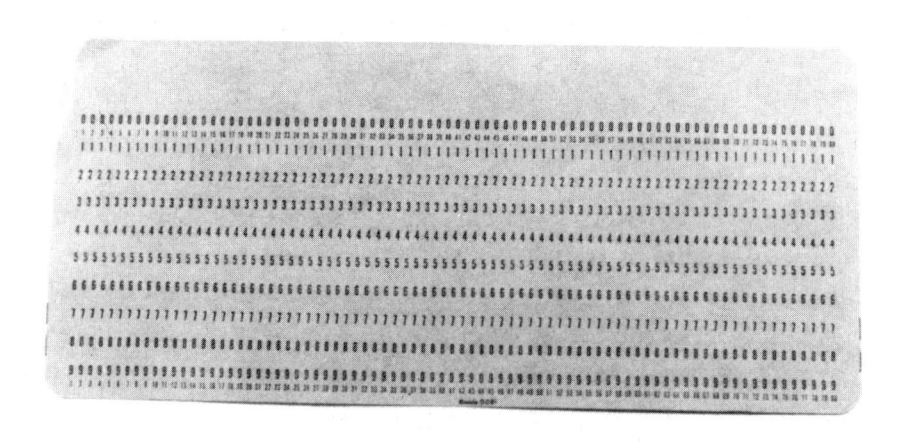

Figure 17-3. A punched card, designed on the same principles Hollerith used to tabulate the 1890 U.S. Census. These cards were the primary method of data entry for computers until the floppy disk replaced them. Some were still in use into the mid-1980s.

were still a primary method of entering information into computers in the 1980s.

Early Computers

The first electronic digital computer was the Atanasoff-Berry Computer (ABC), developed in 1942. The ABC was developed at Iowa State University by Dr. John Atanasoff and his assistant, Clifford Berry. Atanasoff built the computer so his students wouldn't have to take so long in solving long mathematical problems.

By 1944, Howard Aiken, a professor at Harvard University, had developed the Mark I for IBM. The Mark I was really electro-mechanical in that it used thousands of mechanical switches, called relays. When operating, the Mark I supposedly sounded "like a roomful of ladies knitting." One day the Mark I stopped running because a moth was caught in one of the switches. The student who was working in the laboratory wrote "got d'bug out" in his log. The term debug, meaning to correct a problem, was born.

Dr. John Mauchly from the University of Pennsylvania, with the help of a graduate student, J. Presper Eckert, Jr., developed the ENIAC - Electronic Numerical Integrator And Computer in 1946. The ENIAC was built as a wartime secret to figure trajectory tables for the Army. The first large-scale electronic digital computer, it could do 5,000 additions/minute and 500 multiplications/minute. The ENIAC weighed 30 tons and occupied 1,500 square feet of floor space. It contained over 18,000 vacuum tubes requiring a huge amount of electricity and generating a large amount of heat. It operated, on average, seven minutes before a vacuum tube would burn out. The ENIAC was easily the most famous of the early computers and is known as the first general-purpose computer.

The Computer Generations

The UNIVAC, Universal Automatic Computer, was developed in 1951. It was the first commercially available computer. The development of the UNIVAC began what is now called the first generation of computers. First generation computers used vacuum tubes for processing and memory and magnetic tapes for storage.

The first generation also saw the development of the first computer programming language to use English commands—*FORTRAN*, an acronym for Formula Translation. Before FORTRAN was developed in 1957, computer programmers used machine language (zeroes and ones, open or closed). The number of computer programs developed increased dramatically.

In 1959, the *transistor* was invented. This very important invention marked the beginning of the second generation of computers. The transistor replaced vacuum tubes as the primary electronic component in computers. This change allowed computers to be smaller, faster, and to produce less heat. All vacuum tubes in computers were replaced within a few years. The IBM 1401 and Honeywell 400 were both second generation computers.

The second generation lasted until around 1965, when the *microchip* was developed. The microchip contained minute integrated circuits on a silicon chip often smaller than a dime. This made third generation computers smaller and faster. Silicon was used because it is a *semi-conductor*. Today, these chips contain the computer's memory and microprocessor. The IBM 360 was the dominant machine of the generation. The "360" was used to represent the degrees in a full circle, because it was supposed to meet everyone's needs.

The development of very large scale integration (VLSI), led to the fourth generation of computers. VLSI made it possible to put a much larger number of circuits in a smaller space. The IBM 370 was the first computer to make use of this technology.

Computers of the fourth generation are commonly called microcom-

Figure 17-4. A 5.25-inch floppy disk.

PRINCIPLES OF COMPUTER OPERATION 9

Figure 17-5. A compact disk (CD). These optical disks are used to store large quantities of information but are still portable. This disk contains over 30,000 files.

puters, personal computers, or desktop computers because of their small size. The fourth generation continues today, with most experts waiting for computers to exhibit true *artificial intelligence* before signalling the start of a fifth generation. The fourth generation has improved since its inception in the early 1970s. The *floppy disk*, developed in 1973, made it possible to transfer data to and from computers without bulky tapes.

The development of Compact Disk—Read Only Memory (CD-ROM) technology around 1983 made it possible for computer users to have portable storage of large amounts of information. One CD can hold 680 *megabytes* of information, or about 275,000 typed pages of text. This amount is the equivalent of about 1,800 floppy disks. As an example, the entire U.S. telephone directory can be stored on two CDs.

The CD has become popular for storage of large *graphics* files, allowing for interactive video/hypermedia computing. They are also used to hold large databases, allowing for one-time searches of huge quantities of information. Other developments, such as graphical user interfaces and hard disks will be discussed later in this chapter.

Early Microcomputers

The Altair 8800, introduced in 1975, was a do-it-yourself kit designed for electronic buffs. It is generally recognized as the first microcomputer.

In 1976, Steven Jobs and Stephen Wozniak built the Apple computer in their garage. It was mounted on a piece of plywood and built out of spare parts. According to legend, Jobs and Wozniak offered their computer to Hewlett-Packard, the company they worked for at the time, but management there didn't foresee a market for microcomputers. They went on to develop Apple Computer Corporation, one of the largest computer companies in the world.

Radio Shack introduced its TRS 80 series in 1977. Through the mid-1980s, Radio Shack used their system of retail centers to sell many of their TRS-80 microcomputers. It was one of the top early microcomputers.

In 1981, IBM introduced the IBMpc (personal computer). IBM took advantage of its credibility with larger computers to sell many of these poorly-developed models, but did improve upon it with the IBMpc XT and IBMpc AT. IBM and its clones (IBM-compatibles) captured much of the

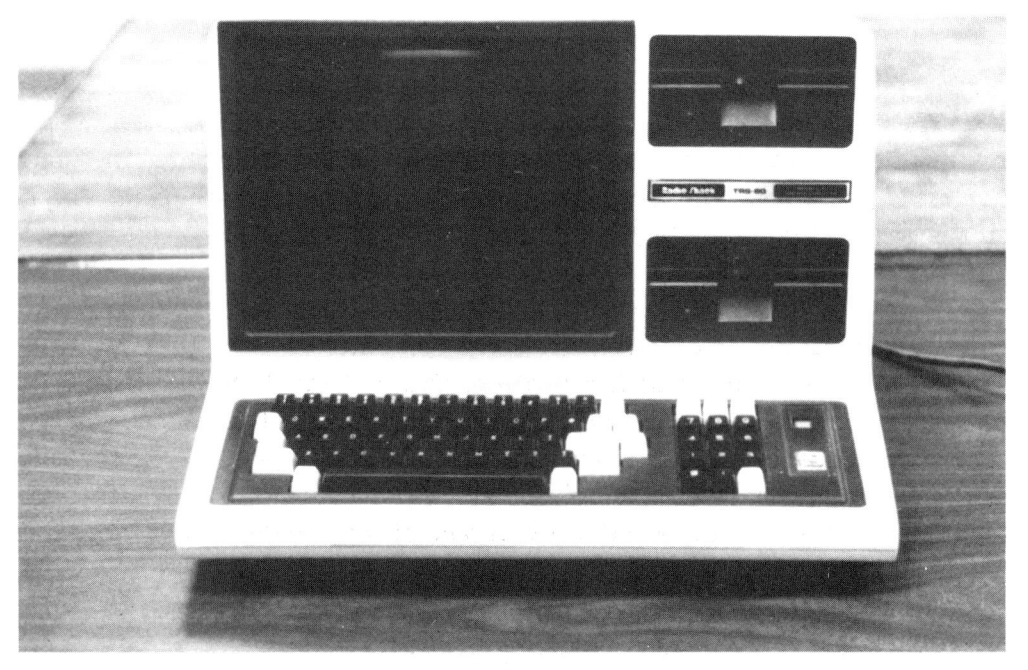

Figure 17-6. A Radio Shack TRS-80 Microcomputer.

PRINCIPLES OF COMPUTER OPERATION

Figure 17-7. The original IBMpc. IBM used this machine to vault to the top in microcomputer sales in the 1980s.

market during the latter half of the 1980s due to agreements with Microsoft Corporation. Microsoft writes programs to run on these computers and has a policy of letting anyone write and market programs using its operating system (MS-DOS).

IBM contracted with Bill Gates, a college dropout, to develop the operating system for the IBMpc. Gates is now chairman of the board of Microsoft Corporation. With a net worth of over seven billion dollars, Gates is the richest man in America.

In 1984, the Apple company developed the Macintosh. With this computer, they introduced the *graphical user interface* (GUI, pronounced gooey), touted for its "user-friendliness" or ease of operation and learning. This interface is characterized by the use of a *mouse* to access its pull-down menus and *icons* (symbols representing commands) as opposed to DOS's command line and character-based user interface. Microsoft soon copied the idea and developed its own GUI program—Microsoft Windows.

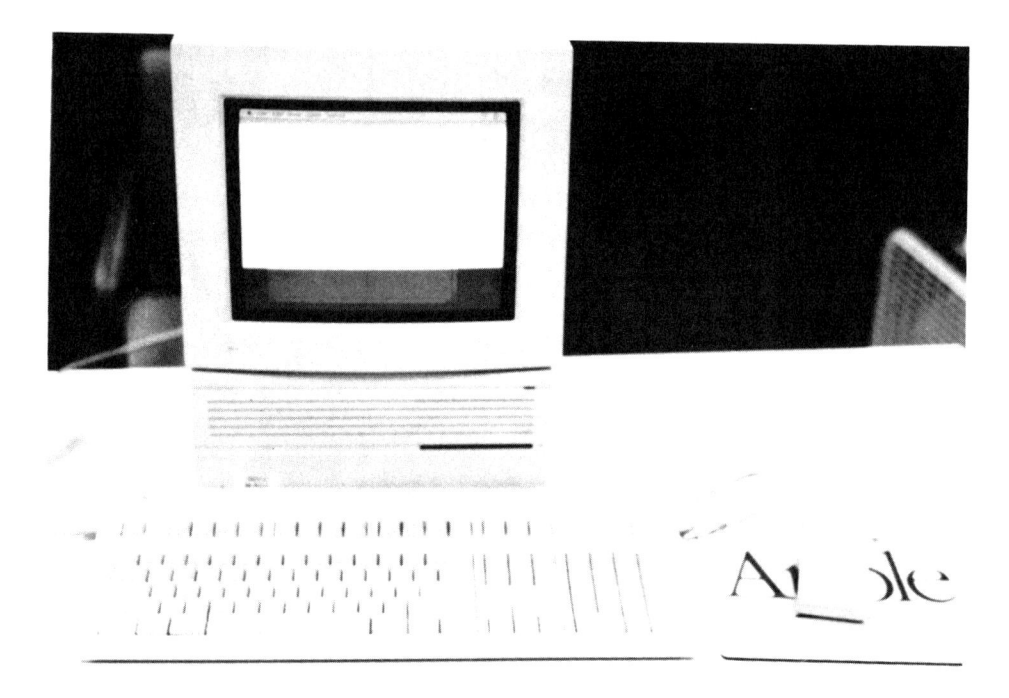

Figure 17-8. An Apple Macintosh. Note the "mouse" to the right of the keyboard. It is used to input commands using on-screen icons.

HOW COMPUTERS WORK

Computers use a series of microchips to translate electrical impulses into information that is meaningful to the user. The process involves microchips that serve as the computer's *memory* and as the *microprocessor* (the "brains" of the computer).

BINARY LOGIC

The basic process used by computers is *binary logic*. Binary, meaning two, is the number of distinctions the computer can make on the signals it receives. When the computer's microprocessor receives a signal, it can only determine whether it is on or off, open or closed. The concept is similar to the electrical circuit behind a light switch—the light is off when the circuit is open and on when it is closed. With computers, 0 represents open and 1 closed.

One signal, open or closed, is called a *bit*, short for binary digit. This amount of information is not very useful in itself, so early computer engineers developed a system whereby a series of bits could be used to represent certain information. They determined that a combination of eight bits could be used to represent all of the characters, such as letters, numbers, and other symbols, needed to program a computer. Collectively, these eight bits are called a *byte*.

A standard code was developed for each particular character. Every computer comes with a built-in ability to translate these characters into information the computer can use. For example, when a computer user types an "A" the computer translates this into *machine language*, which for A is 01000001, then it translates it back to the "A" that shows up on the monitor. The standard codes are called ASCII (pronounced ask-key), which stands for American Standard Code for Information Interchange.

MEMORY

The usefulness of a computer is determined to a large extent by the amount of memory it contains. Memory is the place where information is held so it can be accessed quickly by the computer for the purpose of manipulating the information or executing the program. The two types of memory are Read Only Memory (ROM) and Random Access Memory (RAM).

Read Only Memory (ROM)

ROM is permanent memory which is built into the computer. It is called non-volatile because it remains in the computer when the power (electricity) is turned off. ROM is expensive, so it only contains information that is fundamental to the operation of the computer. These operations include the BIOS (basic input/output system) which allows the computer to relate to peripherals, ASCII codes, and the basic mathematical functions. The ROM resides in microchips in the system unit. These chips can only be read, not written to or altered, once they leave the factory.

Random Access Memory (RAM)

RAM is temporary memory that is used to receive information and execute programs. A program is usually placed in RAM from a disk and resides in RAM until it is replaced by a new program. RAM is volatile—when you turn off the computer, it is erased. Any information in RAM that needs to be saved so it can be used again must be stored on a disk or some other

Figure 17-9. A microchip which is used for RAM memory.

device so it can be loaded into the RAM again at a later time. The random access label comes from the fact that the computer can call up information from any location in memory in the same access time, regardless of which memory location was accessed previously.

Memory Capacity

The memory capacity of a computer is determined by the number of bytes that will fit into the RAM at any one time. Capacity is usually listed in K, or *kilobytes*. One kilobyte is equal to approximately 1,000 bytes (actually, it is 1,024 bytes, or 2 to the 10th power). New computers often list RAM capacity in MB, or *megabytes*. One megabyte is equal to 1,024 K or 2^{20}. As an example, a computer with 1MB of RAM can hold 1,048,576 bytes at one time. A *gigabyte* is equal to 1,024MB or 2^{30}. Some large computers will list their storage space in gigabytes.

Both the program being executed and the *data* it uses must be stored in RAM at the same time. Also, the *operating system* takes part of the RAM, so a computer with 640K will not be able to use all 640K for a program and data. The minimum requirement for most programs today is 640K of RAM, with some programs recommending or requiring 2MB to 4MB. New computers regularly have 2MB to 4MB, and most can be expanded to 16MB. The ability to expand is very important as new programs are developed that require increasing amounts of memory.

PARTS OF A COMPUTER

At first glance, the computer may resemble a television sitting on a box, attached to a typewriter. In fact, these earlier machines did influence the modern look of a computer. The four primary *hardware* components

PRINCIPLES OF COMPUTER OPERATION 15

of a computer system are the system unit, the monitor, the keyboard, and a printer.

SYSTEM UNIT

The system unit contains the actual computer, with other parts considered *peripherals*. Normally, a system unit includes the power supply unit, the motherboard, the disk drives, and the input/output ports.

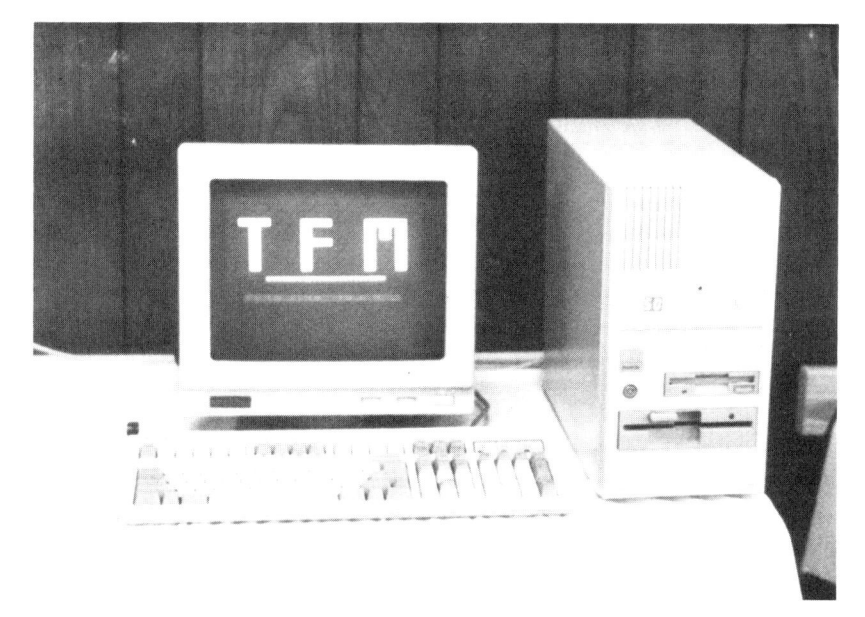

Figure 17-10. A system unit built in a mini-tower format.

Power Supply

The power supply unit simply transmits electricity from the power source to the motherboard to allow the computer to operate. In some countries, the power supply must also have an adapter to convert from direct current (DC) to alternating current (AC) electricity.

The power supply is usually sealed at the factory. Amateurs should not attempt to repair this part or even to try to open it. If a problem occurs, the user should call a technician for any necessary repairs.

A constant source of power is important for the operation of a computer. If the power goes off, all information in RAM is lost. The computer user

Figure 17-11. An Uninterrupted Power Source (UPS), with a battery backup to circumvent a power loss.

should save his or her data often to reduce the catastrophic effects of power loss. Some computer systems, especially networked systems, use an uninterrupted power source (UPS) to avoid data loss. The power source runs through the UPS before entering the computer system. The UPS contains batteries to keep the system running for a period of time with no power coming in from the power source.

A surge protector is another good investment for the computer user. It will not protect against power loss, but will keep surges in electricity from harming the system or turning the system off momentarily.

Motherboard

The motherboard contains the primary microchips that operate the computer; the microprocessor, the memory chips (ROM and RAM) and expansion slots. It contains connections to every part attached to the computer.

Microprocessor. The microprocessor is the chip which processes all of the information given to the computer and transmits all of the output. It is sometimes called a central processing unit (CPU). The microprocessor's

PRINCIPLES OF COMPUTER OPERATION

operating speed and rate of handling information determines the speed at which the computer operates. Most IBM and compatible computers use microprocessors developed by the Intel corporation.

The original IBMpc microprocessor, the Intel 8086, was very slow by today's standards. It handled 16 bits of information at a time and operated at a clock speed of 4.77 megahertz (mHz). The 8086 had 29,000 transistors on one chip and could perform at .33 million instructions per second (*MIPS*).

The 8086 was followed by the 8088, the 80286, the 80386, and the 486; there were several varieties of each. The latest Intel microprocessor, the Pentium, has a 64-bit data handling capability and operates at 60 to 100 mHz. Introduced in 1993, the early versions have 3.1 million transistors on one chip and operate at 112 MIPS. Later versions will be even more powerful.

Other companies developing microprocessors for IBM-compatible computers include Advanced Micro Devices (AMD), NEC, and Cyrix. Apple Macintosh microprocessors are made by Motorola.

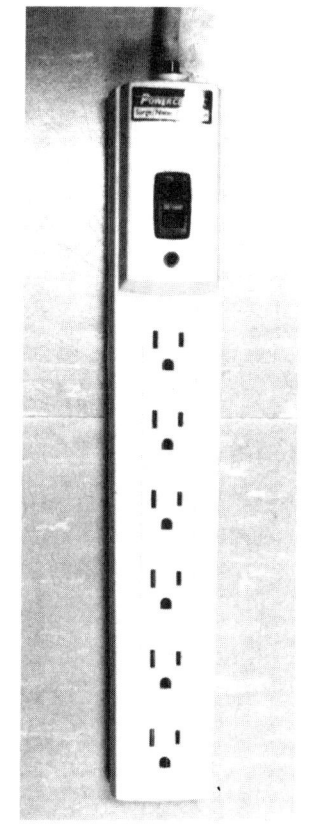

Figure 17-12. A surge protector protects against loss of data due to surges in electrical power.

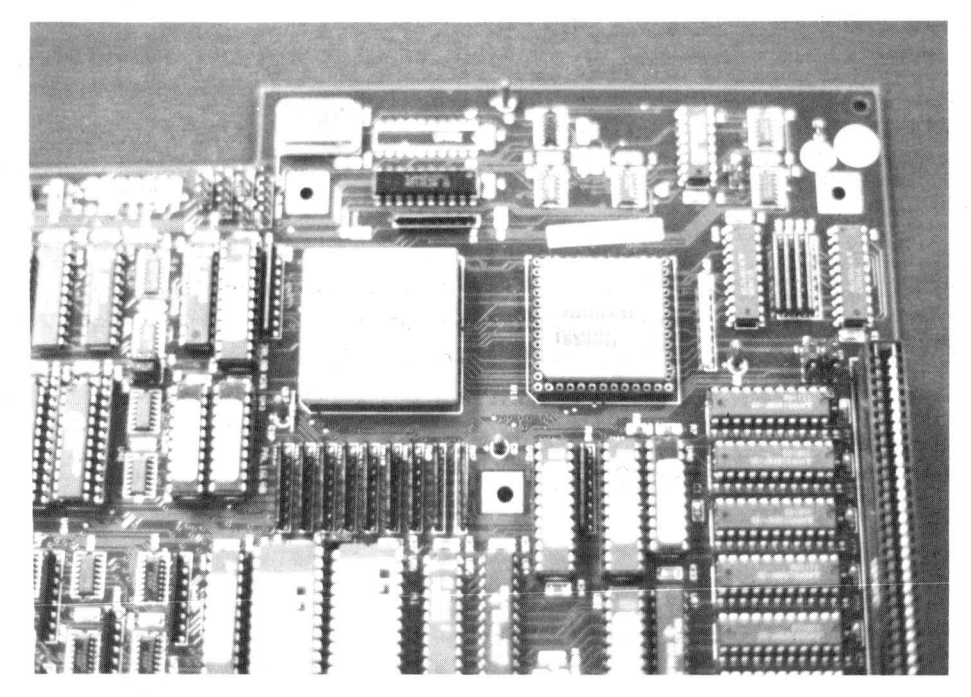

Figure 17-13. A microprocessor and math coprocessor plugged into a motherboard.

Early microprocessors may be enhanced in their ability to handle math-intensive operations by adding a *math coprocessor*. The microprocessor will divide calculations between itself and the math coprocessor, almost doubling the processing speed of drafting programs and other math-intensive software.

Memory. The computer's RAM and ROM chips are also located on the motherboard. They determine the amount of information a computer can hold for immediate access at any one time. The amount of ROM varies from computer to computer, but all hold basic input/output information, arithmetic functions, and logical operators such as equal to and greater than/less than functions. Because ROM is read-only, the user cannot use it for programs or data. The operating system, various *terminate-and-stay-resident (TSR)* programs, *application* programs, and data all reside in RAM. The amount of RAM in computers varies, but as a general rule, the more the better. Most new computers come with 4MB or 8MB of RAM, but can be expanded to 16MB or more.

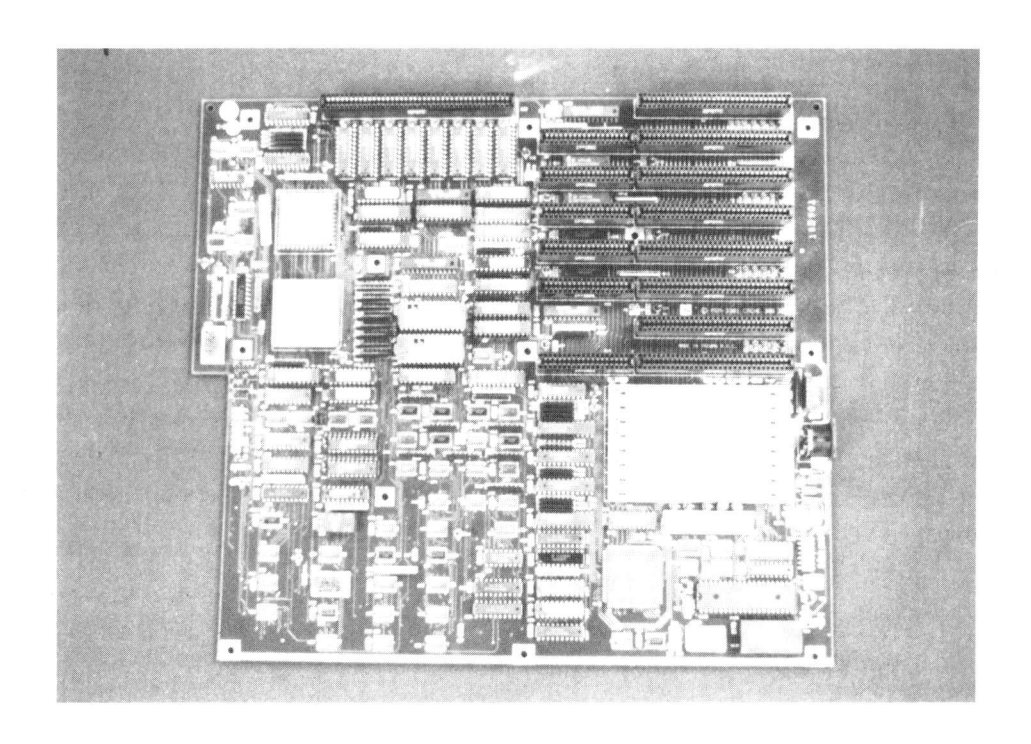

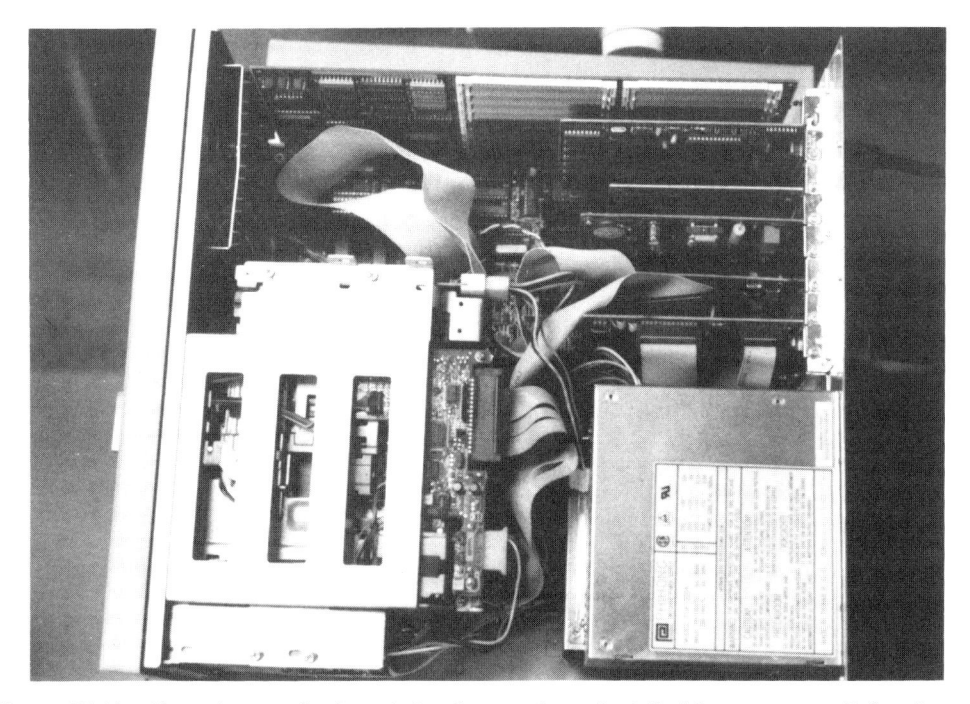

Figure 17-14. Above is a motherboard that has not been installed into a system. Below is a small footprint computer with the cover removed. Note how many expansion slots are filled with cards.

Expansion Slots. Expansion slots allow the motherboard to be expanded to connect to other devices or for additional memory. A card is inserted into a slot that connects it to the motherboard. The card may have a port to connect it to another device out of the back of the computer. Most computers come with eight or more expansion slots, but several of these will already be filled. A computer buyer should determine how many expansion slots are available before buying a new computer. (Most computer descriptions will give total and available expansion slots.)

A small footprint IBM compatible computer (which takes up less desk space) with a bus modem, a mouse, and an internal modem that is connected to a network may have only one expansion slot available, even though it started with eight.

- One expansion slot is usually taken for output—the parallel port. This is usually used to connect the computer with a printer, but may be used for other output devices such as a plotter.
- Computers with a small footprint will often use a slot for some of the memory because of the small size of the motherboard.

Figure 17-15. A mouse and mousepad.

Figure 17-16. An internal modem. The modem plugs into an expansion slot on the motherboard.

- One expansion slot is usually taken for the graphics card, which has an output port connecting it to the monitor.
- One expansion slot is used for a connection to the various disk drives a computer may have.
- If the computer has a bus mouse, it will take up an expansion slot. The card will have an input port to connect the mouse.
- An internal modem or FAX/modem card will also use a slot. It will have ports for incoming and outgoing telephone lines.
- If the computer is connected to a network or a printer-sharing device, these may need to use an expansion slot.

Disk Drives

The primary location for storing information for the computer to use is on disks. The data is stored as electronic impulses on magnetic disk material. The device used to read and write information using disks is called a *disk drive*.

Most computers have a least one, usually two, floppy disk drives and one hard disk drive. Some computers may also have a CD-ROM drive or a tape drive (used almost exclusively to *backup* a hard disk in case of failure). These are located in drive bays in the system unit. A personal computer will usually have three full-height drive bays and one or two half-height drive bays. The full height bays are used for 5.25-inch drives or hard disk drives. The half-height bays are usually used for 3.5-inch drives.

Floppy disks must be formatted in order to store information. Formatting maps the disk into sectors and tracks that the drive uses to locate information on the disk. The number of sectors and tracks depends on the size and type of the disk and determines the amount of information which can be stored on the disk.

Disk drives contain a read/write head which simultaneously reads or writes to both sides of a disk. This makes the disk system much faster than the tape systems which were used in the earlier generations of computers, because the tapes must access information sequentially. The tape drive may have to rewind almost the entire length of the tape to get to a certain piece of information.

While the disk spins on the drive, the read/write heads move from outside to inside to get the information quickly. Because the disk spins faster than the read/write heads move, information is stored from outside to inside on both sides at the same time. This allows for the fastest access to information.

Floppy Disk Drives. Floppy disk drives generally come in two sizes, 5.25-inch and 3.5-inch (sometimes called microdiskettes). Older computers may have double-density disk drives, which will only read and write on double-density disks. Most newer computers come with high-density disk drives, which can read and write on high- or double-density disks. When purchasing a new computer, most buyers opt for one disk drive in each size. This allows them to use either 5.25-inch disks or 3.5-inch disks.

Floppy disks may contain space for 360K of information (5.25-inch, double-density) or up to 2.88MB of information (3.5-inch, super high-density). Table 17-1 contains an explanation of how the various floppy disks store information.

Special care must be taken to avoid damage to floppy disks. With improper handling, data may be lost or destroyed and impossible to recover. The smaller 3.5-inch disks are sturdier and have a sliding cover for the

PRINCIPLES OF COMPUTER OPERATION

Table 17-1
FLOPPY DISK SIZES AND STORAGE CAPABILITIES

Size	Density	Tracks/Side	Sectors[1]/Track	Storage Capacity
5.25"	Double	40	9	360K
5.25"	High	80	15	1.2MB
3.5"	Double	80	9	720K
3.5"	High	80	18	1.44MB
3.5"	Super-High	80	36	2.88MB

[1] Each sector contains 512 bytes.

write/protect notch, but still can be damaged. Some special concerns when handling floppy disks:

- Keep disks away from magnets or magnetized objects (such as a ringing telephone or magnetized screwdriver). Magnetized objects can erase all of the electronic impulses.

- Don't smoke around the disk. Smoke particles are larger in size than the distance between the read/write head and the disk surface. The same is true for food and drink particles.

- Never touch the actual disk surface. Body oils can damage the surface.

- Use a soft-pointed pen, such as a felt tip, to label disks.

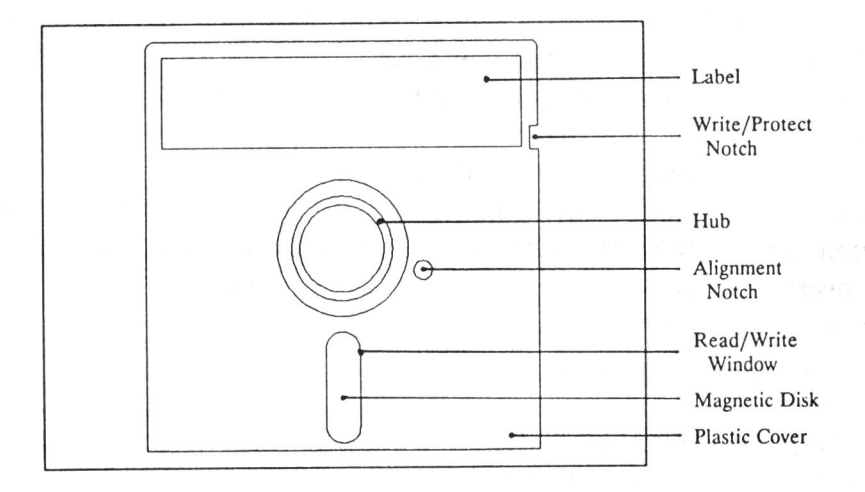

Figure 17-17. Parts of a floppy disk.

Figure 17-18. A 3.5-inch floppy disk.

Hard Disk Drives. Most computers have a hard disk drive with a fixed *hard disk*. This means that the disk and drive are not portable; they stay in the computer at all times. Computers running new programs must have a hard disk drive, because the operating system and programs will not fit on a floppy disk. Many users also store most of their *data files* on the hard disk.

Hard disk drives work the same way as floppy disk drives, but the hard disk usually has several platters. The platters are made of aluminum or ceramic and coated with a magnetic surface material. The disk drive has two read/write heads for each platter on the hard disk. The read/write heads work like a phonograph needle, except they don't actually touch the hard disk, although they do come very close. A smoke particle is larger in diameter than this distance, which is why a computer user shouldn't smoke around the computer. (Of course, smoking is also bad for the health of the user and others who work around that person.)

The first hard disks contained space for 10MB to 30MB of information.

PRINCIPLES OF COMPUTER OPERATION

Today's personal computers may have up to several gigabytes of storage space. The most common hard disks, however, have from 120MB to 500MB of storage space.

Hard disks usually are located in one of the available disk drive bays. Some hard disks, however, are located on a card which fills an expansion slot. These are referred to as hard cards. Some computers make use of portable hard disks, called Bernoulli disks. These disks may be used for primary storage or to backup the fixed hard disk.

A disk cache is used on newer computers to speed up operations that are slowed down by accessing the hard disk drive. The disk cache keeps copies of recently accessed sectors on the hard disk in a reserved area of memory called cache RAM. When the program tells the computer to access the hard drive, the cache program first goes to cache RAM to see if the information needed is located there. If the information is in cache RAM, the operation occurs faster because accessing RAM is much faster than accessing a hard disk.

Figure 17-19. A hard disk drive with the cover removed. Note the three platters and read/write heads for each exposed side.

Other Storage Devices. The other two primary storage devices used on computers are the tape drive and the CD-ROM drive.

The tape drive is used primarily to backup a hard drive as insurance against loss of data if disk failure occurs. It may use cassettes or tape cartridges. Newer tape drives may use 8mm tapes similar to the ones used in video cameras. These tapes may hold several gigabytes of information. An internal tape drive will be located in one of the disk drive bays. An external tape drive will be attached to the computer through the serial port.

The CD-ROM drive reads the optical compact disks (CD) mentioned earlier in this chapter. The CD can store digitized text, graphics images, and sound. Like the tape drive, it may be internal or external.

Input/Output Ports

The system unit may have several input/output ports to connect pe-

Figure 17-20. Tape drives used to backup files on a network file server. The one on the left uses an 8mm cartridge and the one on the right a 1/4-inch cassette.

PRINCIPLES OF COMPUTER OPERATION

Figure 17-21. An external CD-ROM drive, connected to the computer's serial port.

ripherals to the computer. The two standard ports are the serial port and the parallel port.

Serial ports. The serial port is an asynchronous communications port. A mouse, an external modem, or a character printer can be connected to the computer through this port. The term serial is used because data is transferred one bit at a time over a single line.

Serial ports use an RS-232 interface, which is a standard that was developed to allow devices to be used interchangeably on different computers. The computer addresses the serial ports as COM1, COM2, etc. Although most use fewer, one computer can address up to four COM ports.

Parallel ports. Parallel ports are normally connected to a printer of some type. The term parallel means that bits of data are transferred over more than one line at a time. This process makes parallel ports faster than serial ports. The parallel port is often connected to the motherboard with a controller card, thereby using one of the expansion slots.

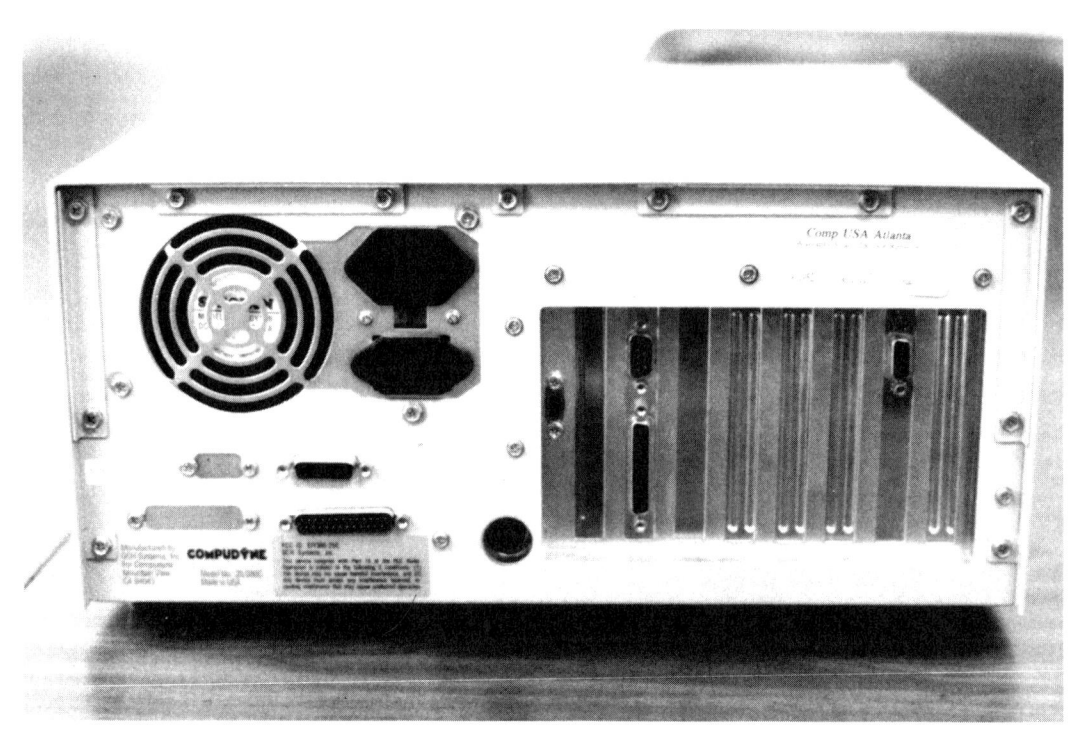

Figure 17-22. The back of a computer. Note the input/output ports available for various connections.

MONITOR

A computer monitor, or screen, is an output device which allows the computer user to see the results of processing. For example, when the user types in a command, the command shows up on the screen so the user can see what was typed. When the command is processed, the results of the process show up to let the user know what occurred or why the command did not work.

The four primary types of monitors or screens found on personal computers are the *cathode ray tube (CRT)*, the electroluminescent (EL) display, the *liquid crystal display (LCD)*, and the gas-plasma display.

Cathode Ray Tube (CRT)

The most common type of monitor for personal computers is the CRT. The CRT is based on the same technology found in a television set—some can even be used to receive television signals. They usually display 25 lines

PRINCIPLES OF COMPUTER OPERATION

of text with each line 80 characters long. A 12- to 15-inch size, measured diagonally, is the most common, although some are larger.

The surface of a CRT screen is coated with a phosphorescent coating. When the monitor receives a signal from the computer, an electron gun generates an electronic beam on the phosphors, which causes them to glow. Each phosphor glows for a short period of time and then must be refreshed. The refresh rate is one of the factors that determines how easy the monitor is to look at for extended periods of time. A standard refresh rate to keep the image from flickering is 60 times per second, but superior monitors may be much faster.

Color CRT monitors contain red, green, and blue electron guns. Each color produced on the monitor is some combination of these colors at various strengths. For example, white is the result when all three guns are turned on high. Various shades of gray occur when all three are reduced by the same amount.

Monitors with graphic capabilities have bitmapped displays. The graphics standard information is located on a graphics card in an expansion slot. The screen has a number of dots that are illuminated to create a graphic or text. Each dot that can be illuminated is called a *pixel*. The number and size of the pixels determines the sharpness of the image produced, also called the *resolution*. The size of the pixel is called the *dot pitch*. A smaller dot pitch and higher resolution produce a sharper picture. Table 17-2 contains a summary of the various monitor standards and resolutions.

Table 17-2
VIDEO DISPLAY STANDARDS

Standard	Description	Year	Resolution	Pixels
MDA	Monochrome Display Adapter	1981	720 × 350	252,000
CGA	Color Graphics Adapter	1981	640 × 200	128,000
EGA	Enhanced Graphics Adapter	1984	640 × 350	224,000
VGA	Video Graphics Array	1987	640 × 480	307,200
SuperVGA	Extended VGA/VGA Plus	1988	800 × 600	480,000
			1024 × 768	786,432

Source: *PC Glossary*, Disston Ridge, Inc.

Electroluminescent (EL) Display

The EL is often called a monochrome monitor, meaning one color. It is the most reliable type of display and is found on many computers in operation. Most EL monitors do not have graphics capabilities, but are excellent for text, providing a sharp image.

EL monitors have a phosphor film between a reflective back film and a transparent front film. A grid of electrodes contains the pixels which can be lighted individually. These monitors usually emit an amber or green color on a black background, which is easy on the eyes of the user.

Liquid Crystal Display (LCD)

LCDs use a liquid crystal material sandwiched between two pieces of glass to form the screen. The image is activated by polarizers and an external light source. LCD's are commonly found on portable, especially notebook, computers because they have a flat screen that is not bulky when compared to a CRT. The same technology is found in wristwatches, clocks, microwaves, and many other devices.

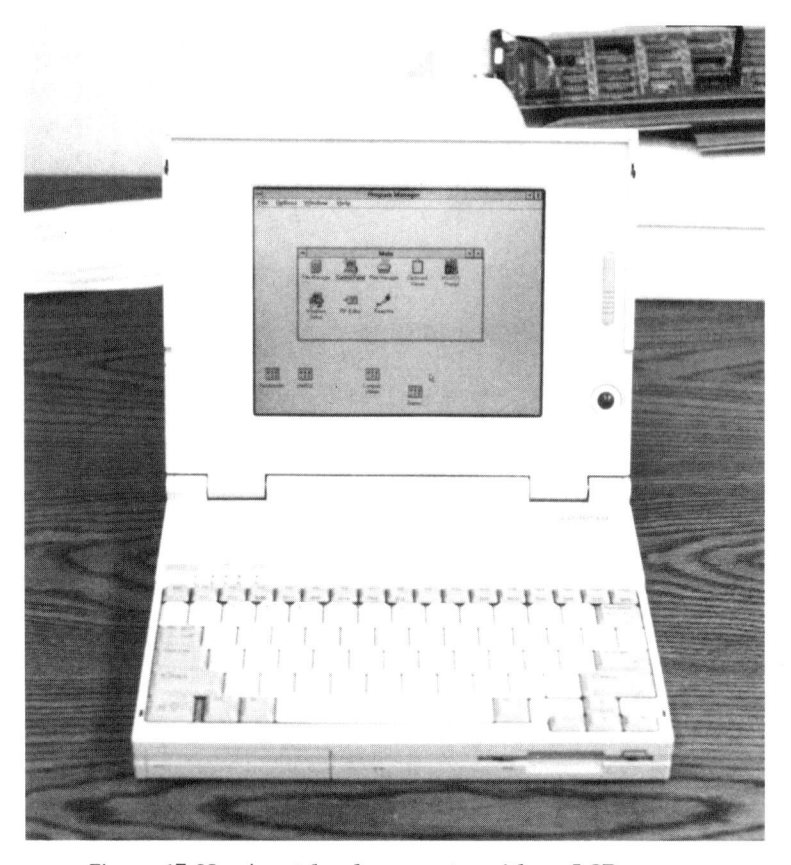

Figure 17-23. A notebook computer with an LCD screen.

On the first LCDs, the image could only be viewed from directly in front of the screen. Newer versions have backlighting and advanced electronic signals to produce a much better image than the early models. LCD's are now available in color and monochrome versions.

Gas-Plasma Display

Gas-plasma displays use a trapped gas, usually neon or argon/neon, which lights up when electricity is applied to it. The electricity causes the individual pixels to glow an orange-red color.

As a monochrome screen, the gas-plasma has a sharper image than the LCD. Like the LCD, it is a flat screen and is used almost exclusively on portable notebook computers.

KEYBOARD

The keyboard is the computer user's most common input device. Along with a mouse, it is the means by which the user can tell the computer what to do. The keyboard is the detachable device that looks similar to the keypad of a typewriter. The primary difference is that the computer keyboard contains some specialized keys for use with computer *software*.

Like a typewriter, the computer contains the letter and number keys, special character keys, a shift key, a caps lock key, a backspace key, a tab key, and the spacebar. On a computer keyboard, however, the return key is replaced with an enter key, which signals to the computer that a command is complete. The computer keyboard also contains 10 or 12 *function keys*, a control key, an alternate key, and *cursor control keys*.

The cursor control keys include arrows and other keys used to move the *cursor* around. The other cursor control keys are the home key, the end key, insert and delete keys, and the page up and page down keys. The cursor control keys may be located on a *numeric keypad*, found on most keyboards, with a switch to toggle the keys between their different functions. The keyboard may have dedicated cursor control keys in addition to those on the numeric keypad. On some notebook computers, the computer will not have a number pad, and the cursor control keys will be located on the letter keys, usually under the right hand.

Most keyboards use the standard typewriter configuration for the letter keys, called a QWERTY keyboard. (The first five letters in the top row are q, w, e, r, t, and y.) The QWERTY keyboard was first developed to slow down fast typists, because they were jamming the keys on mechanical

Figure 17-24. A 101-key, QWERTY computer keyboard. Note the numeric keypad on the far right and the dedicated cursor control keys between the letter keys and the numeric keypad.

typewriters. Some other configurations include the Dvorak, Maltron, and AZERTY keyboards.

PRINTER

The printer is a common output device that provides the computer user with a hard copy of information. This protects the user by providing another copy of data, in case the original is lost or erased from the computer's storage. It also provides a portable copy that can be shared with those who do not have a computer or appropriate software to read the electronic copy.

The four major types of printers are daisy wheel, dot matrix, laser, and inkjet.

Daisy Wheel Printer

The *daisy wheel printer* is an impact printer based on technology available in many typewriters. A circular disk (the wheel) has two characters on each individual spoke. To print a letter, the printer strikes the individual spoke against an ink ribbon. This action results in a complete letter printed on the paper.

The daisy wheel is a letter-quality printer. This means that the output is comparable to a typewriter's output—the letters are fully-formed and easy to read.

Daisy wheel printers commonly use either a tractor-feed mechanism to move continuous form paper through the printer or a single sheet feeder. Some use a friction feed system, much like that of a typewriter.

PRINCIPLES OF COMPUTER OPERATION 33

Daisy wheel printers may come in standard carriage, which handles a page 8.5 inches wide, or wide carriage, which can handle a 14-inch wide page. Standard paper can hold about 80 characters on one line, while wide paper can hold about 132 characters.

Dot Matrix Printer

The *dot matrix printer*, the most common printer for individuals, prints characters composed of dots. It prints characters one at a time by pressing wires, also called pins, against an ink ribbon and onto the paper. The wires are arranged in a rectangle.

Dot matrix printers come in two major varieties, 9-pin and 24-pin. The 24-pin printers are more expensive and produce a higher quality output. Most dot matrix printers are bidirectional, meaning that they can print

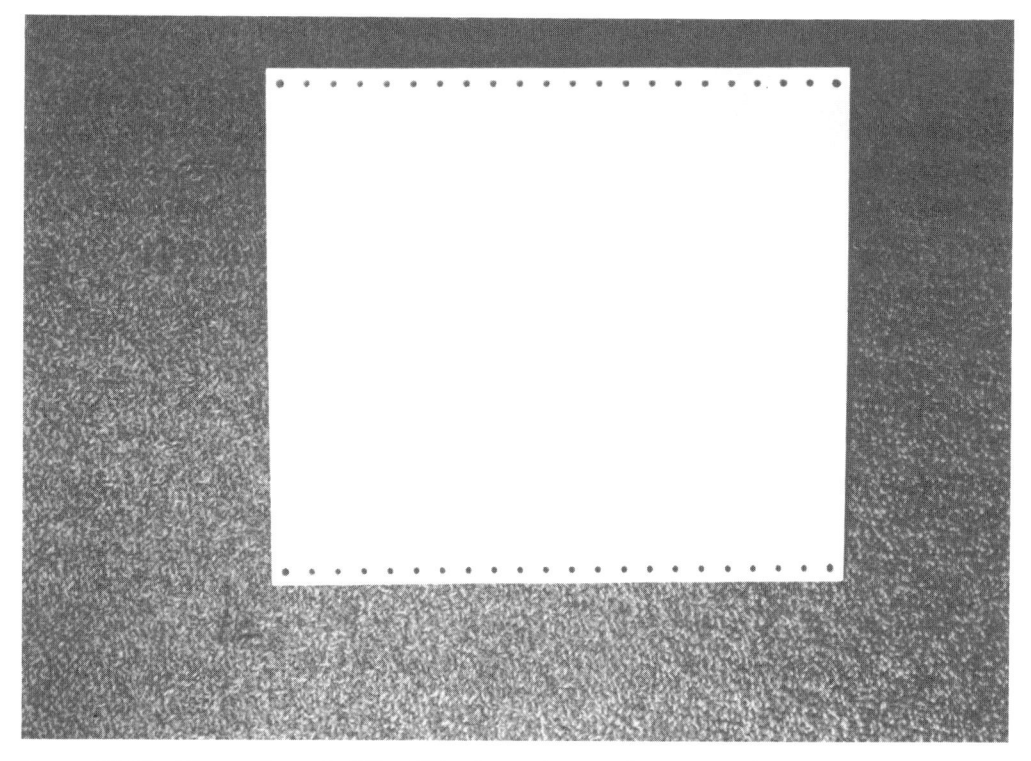

Figure 17-25. Nine and one-half by 11-inch continuous form paper for use with a computer's printer. Note the perforated edges which allow the printer to use a tractor feed mechanism. When the perforated edges are removed, the paper is a standard 8.5 by 11 inches.

Figure 17-26. A 24-pin, dot matrix printer.

lines of text from left to right or right to left, which increases printing speed.

Dot matrix printers usually have two modes, draft and *near letter quality*. In draft mode, dot matrix printers are fast, but the output is of lower quality. In near letter quality mode, the print is improved because each line is printed two times. With a new ribbon, a 24-pin printer is very close to letter quality, and is usually considered acceptable for correspondence.

Dot matrix printers can print medium-to-high quality graphics. The printer uses *bitmapping*, where the computer tells the printer whether or not to print each dot on the page.

Like daisy wheel printers, dot matrix printers commonly use either a tractor-feed mechanism or a single sheet feeder, although some use a friction feed system. Many dot matrix printers can change from one method of feeding paper to another. Also, like daisy wheel printers, dot matrix printers may be standard carriage or wide carriage.

Laser Printer

The *laser printer* prints output one page at a time, rather than one character at a time. It works somewhat like a photocopier, with the image coming from the computer, rather than another piece of paper. When the computer sends a signal to the printer, it focuses a laser beam on its drum. Each position on the drum touched by the laser becomes negatively charged and attracts toner. Then heat and pressure transfer the toner to the paper

PRINCIPLES OF COMPUTER OPERATION 35

and fuse it to the paper's surface. This is why some people refer to the laser as "burning" an image on the paper.

Laser printers are popular because they are very flexible, easily combining letter quality text in various sizes and high quality graphics on one page. Like a dot matrix printer, the laser prints graphics by letting the computer tell it whether or not to print each dot. With a laser, however, many more dots must be bitmapped. Standard resolutions of laser printers

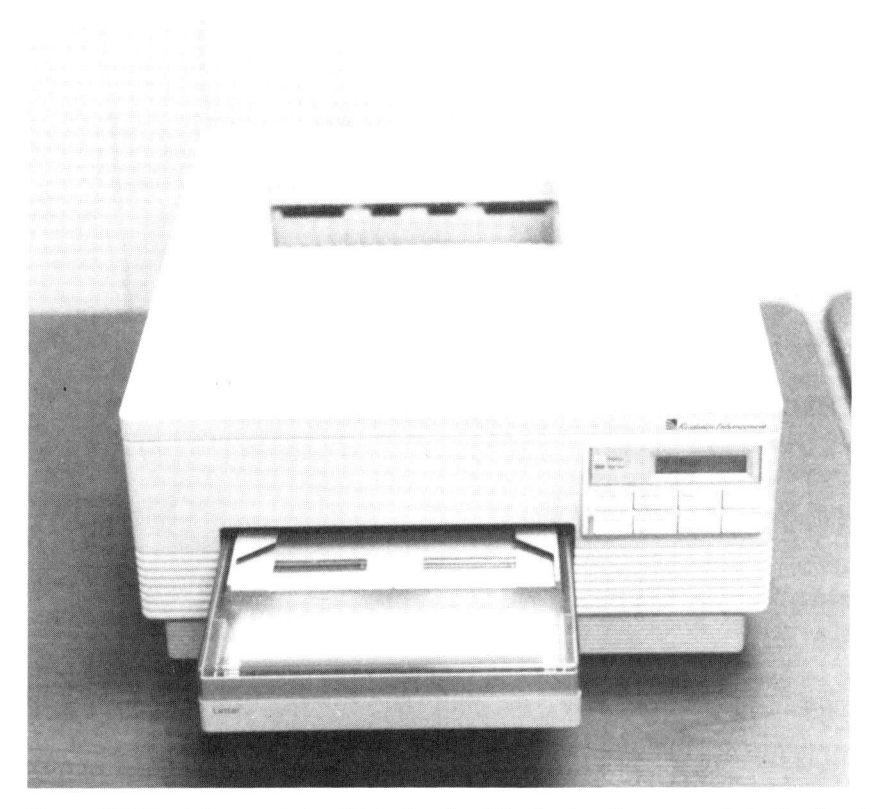

Figure 17-27. A laser printer. Note the sheet feeder loading paper into the front of the machine.

range from 300 dots-per-inch (dpi) to 1200 dpi. Because the entire page is printed at once, the printer must have enough memory to bitmap an entire page. For high-quality graphics, the laser should have at least 2MB of memory and may need more.

Lasers typically use a sheet feeder that handles paper 8.5 inches wide and may hold from 100 to 500 sheets of paper at one time.

Inkjet Printer

Inkjet printers use a spray nozzle to spray ink on the paper to produce a character or graphic. An ink cartridge slides across the page as the paper moves up one line at a time. The ink may be one color or several different colors.

Color inkjet printers are popular with users who do graphics work requiring different colors. Like dot matrix printers, they bitmap the graphic and print it one line at a time. They also can do overhead transparencies in colors.

Inkjet printers can be made very small. This feature makes them popular for use with laptop computers.

Inkjets vary in the quality of their print, but have improved greatly in the 1990s. The output of high-quality inkjet printers is considered letter quality.

COMPUTER SOFTWARE

Software refers to computer programs that allow the computer user to accomplish specific tasks. In general, software can be classified by how the user purchases or obtains the software and by its function.

METHODS OF OBTAINING SOFTWARE

Software may be *freeware, user-supported software,* or *commercial software.* Each has features that make it different from the others.

Freeware

Freeware refers to a freely-distributed software that has been placed in the public domain. This means that the author has not attached a copyright notice to the software and does not retain any legal rights to the program(s). The software may be used, freely copied, and distributed by anyone. While the user is not required to pay a fee to use the program, the author may attach a request for comments about the usefulness of the program to the software.

User-supported Software

With user-supported software, the author retains the copyright. The program can be freely copied and distributed, but the user is expected to register with the author and pay a fee for using the software.

Payment of the user fee gives permission to use the software and typically includes a full set of *documentation* upon registration. The user may also receive the newest version of the program and future updates. With some programs, registered users are offered telephone support for the software and a commission for registering other users.

User-supported software is sometimes referred to as shareware, which is actually a marketing system for software authors. Shareware distribution allows the author to market a program with minimal expense. Many of these software authors belong to ASP, the Association of Shareware Professionals.

Figure 17-28. PC Glossary is a user-supported package containing user terms and definitions. It is available through Disston Ridge, Inc.

Commercial Software

Most successful application programs are commercial software which is copyrighted by the company or person who distributes it. It is illegal to copy these programs, other than to make backup copies for personal use. The software cannot be given to others without the permission of the copyright holder. Unauthorized duplication is known as *software piracy*.

Commercial software companies have enormous amounts of money invested in the development of software. Like book publishers, the copyright is their primary means of protecting that investment. One way they have tried to avoid software piracy is by using copy-protect codes in their

Figure 17-29. Norton Antivirus, a program designed to detect and prevent viruses from infecting a computer system.

software, but computer users found ways around these codes soon after they were developed.

FUNCTIONS OF SOFTWARE

Many types of software are available in today's market, each with its own special capabilities and functions. In general, however, the software types can be classified into six categories: operating systems, word processing, electronic spreadsheets/recordkeeping, database management, graphics, and communications/networking. A brief description of each of these is provided below. The remaining chapters in this text go into more detail about these special functions.

Operating Systems

The *operating system* is the basic software which allows the computer to operate. Other applications run under the operating system platform; it provides a system from which the computer interacts with the software and the user.

The most common types of operating systems are: Microsoft Disk Operating System (MS-DOS), used with older IBM and IBM-compatible computers; System, used with Apple Macintosh computers; OS/2, used with newer IBM computers; and UNIX, used with networked computer systems.

Also included in this category are utilities programs designed to enhance the functions of the operating systems and hardware. These utilities may be user interfaces, menu programs, *virus* protection programs, and file management programs.

Word Processing

Word processing is the most commonly used function of desktop computers. Over 80 percent of the desktop computers in operation are used at least part of the time for word processing.

Word processing is simply using the computer as a "turbocharged" typewriter. The computer allows the user to select different sizes of print, make revisions easily, and saves the user time because documents that need only a few revisions do not have to be entirely retyped. WordPerfect and Microsoft Word are the two most popular word processing programs in use today.

Desktop publishing software is specialized word processing software for people who produce more complex documents combining text and graphics. It combines some of the features of a word processor with a graphical user interface and what-you-see-is-what-you-get (WYSIWYG, pronounced whizzy-wig) placement and formatting. PageMaker and Ventura Publisher are two popular desktop publishing programs.

Spreadsheets/Recordkeeping

The computer's speed and accuracy when performing mathematical functions allows it to be an excellent tool for keeping numerical records. The two primary types of software for this function are electronic spreadsheets and accounting packages.

Electronic spreadsheets are enhanced versions of the paper spreadsheet that has been used in recordkeeping for many years. The spreadsheet, a very flexible program, allows the user to develop his or her own categories, headings, and formulas within the form. Lotus 1-2-3, Quattro Pro, and Microsoft Excel are three very popular electronic spreadsheet programs.

Accounting packages come with ready-to-use recording systems based on general accounting practices. Their primary feature is that the program sets up accounts and ledgers for the user, saving time. Various levels of flexibility are built into the programs, and some have advanced features such as electronic payments, check writing, and tax computation. Quicken is probably the most popular accounting package used today.

Database Management

Database management programs are designed as flexible programs that enable the user to create a database of information about a set of subjects, then have ready access to the information. These programs allow the user to sort, search, choose, update, and report information in a variety of ways.

Database management software is useful for text and numerical information, with tools to manipulate both types. The user decides on what information is needed about each subject, divides the information into logical areas, and inputs the information. DBase, Paradox, and Oracle are popular database management programs.

Graphics

Graphics programs allow the user to generate a variety of charts, graphs,

Sales Grow in Every Region
1989 – 1993

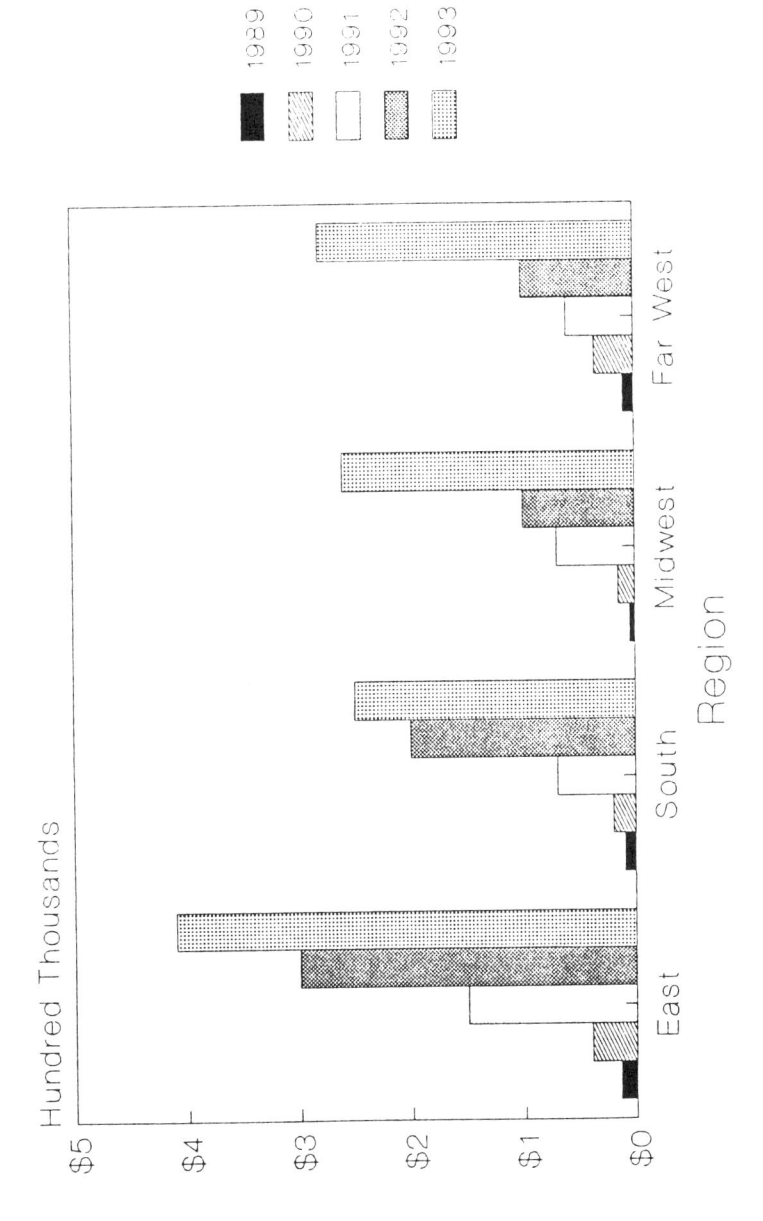

Figure 17-30. Output from a computer graphics program.

and line art. Some of the programs have a set of graphics available; others allow the user to create original drawings; and some do both. Certain graphics programs allow the user to input numbers which are then converted into data graphs, such as bar, pie, or line graphs.

Different graphics programs allow the user to choose from a variety of output forms. The output may be designed to be shown on the computer monitor, projected on a wall screen, printed, plotted, or converted to 35mm slides. Harvard Graphics, Corel Draw, MacPaint, and AutoCAD are popular graphics/drafting programs.

Communications/Networking

The practice of using computers as long-distance communications tools is growing rapidly. In such systems, the personal computer, or a terminal, is connected with a file server, usually through telephone lines. This file server may serve a small area in a local area network (LAN), or a larger area in a wide area network (WAN). Novell, with its NetWare system, controls about 80 percent of the LAN market.

Networking has three primary functions: file transfer, where files are shared between two or more computers; program sharing, where more than one person can use an application; and electronic mail, where messages can be sent electronically in just a few seconds.

BUYING A COMPUTER SYSTEM

One of the many decisions an agribusiness manager has to make is whether or not to buy a computer and, if buying, which computer to purchase. Many factors must be considered, including the type of business, the computer abilities of personnel, future needs, and costs. A good rule is to buy enough computer to accomplish the needed work, while considering future upgrades and prices. When a new microprocessor comes on the market, like the Pentium did in 1993, a smart manager can often get the next lowest technology (in that case a 486DX) at a very good price and still get plenty of computer to do the job.

Below is a brief discussion of the questions a manager should answer before purchasing a new computer system.

1. **What do we need from a computer?** Examples include the ability

to produce letters, billing statements, inventory printouts, financial records, communications, and service manuals.

2. **What applications are necessary to provide what we need?** A word processing program will produce letters and service manuals, but if the company produces several newsletters or brochures, a desktop publishing package may also be needed. For recordkeeping, an electronic spreadsheet may be the choice if someone in the office has experience using one; if not, one of the accounting packages might be more appropriate.

3. **What software do we want to provide our applications?** This choice is often based on the personal preference of the manager and office staff. Two things the manager should keep in mind are the amount of training required and the stability of the software company. The manager wants to avoid lengthy and costly training if possible. Having someone in the office who can use a software package may influence the manager to buy that package. Buying from a stable company means that future service needs and upgrades will be available. One thing the manager might consider is using an *integrated software* package that combines several applications at one price.

4. **What hardware do we need to run the software?** Although it is nice to have state-of-the-art equipment, managers can often save money and get adequate or better service by buying equipment with slightly older technology. Software developments usually run a few years behind the hardware; therefore the programs available may not be able to take advantage of the latest equipment. Of course, if the hardware is needed and will be needed for several years, it pays to buy good equipment.

    The software chosen and amount of data required will determine the specification of the hardware. This includes the type of peripherals, amount of memory, and the size of the hard disk. For example, if Windows-based graphics programs are chosen, a large hard disk and a fast processor are important. If the output needed is 35mm slides, a film recorder must be one of the pieces of hardware.

    Hardware decisions can be a matter of personal preference, sometimes determined by physical characteristics of the user.

5. **What support service is available for the hardware and software?** Service is very important. The software should have toll-free telephone support and good documentation. The hardware should have

486DX33
VESA

- 33 MHz LOCAL BUS INTEL MOTHERBOARD
- 256K CACHE MEMORY
- DESKTOP, MINITOWER, OR SLIM CASE
- 220 WATT POWER SUPPLY
- 4 MEGS OF 70ns MEMORY
- 1.2 MEG FLOPPY
- 1.44 MEG FLOPPY
- 4 MEG IDE LOCAL BUS CACHING CARD
- 130 MEG FAST IDE HARD DRIVE
- 2 SERIAL/1 PARALLEL/1 GAME PORTS
- SVGA 1 MEG LOCAL BUS VIDEO CARD
- SUPER VGA .28dp MONITOR
- 1 YEAR ON-SITE WARRANTY
- MS-DOS VER. 5.0/ WINDOWS 3.1
- NOVELL CERTIFIED

$2479.65

Figure 17-31. A description of a computer for sale in May 1993. Note how the various parts of the system are specified. This allows the buyer to compare prices and features among dealers.

Figure 17-32. Local dealers are often a good choice when buying a computer because of the service they can provide.

a warranty, preferably specifying on-site service, and a good record of reliability. A good policy is to buy locally, when possible. This practice usually ensures good service, but at a slightly higher price.

6. **What money is available to spend on the system?** The manager should avoid making decisions based on price alone, even though it is an important factor. After the other considerations are made, the manager must make a decision that is cost-efficient and still meets the goals of the business. Sometimes the decision is easy. For example, a computer used exclusively for word processing does not need a fast microprocessor and may not need a color monitor. The printer for this system, however, should be of good quality.

The manager's job is to provide office staff—including the manager—with the tools they need to do their work efficiently. This usually means buying computers. The tough decisions are determining what software and hardware to buy and from whom it should be purchased.

CHAPTER QUESTIONS

1. What is a computer program?
2. Describe three early computing devices.
3. What was the importance of the punched card?
4. Describe each of the four generations of computers.
5. What two inventions led to the development of the microcomputer?
6. When was the first microcomputer developed?
7. When was the IBMpc first introduced?
8. What is binary logic?
9. Explain the terms ROM and RAM.
10. Briefly describe the parts of a computer.
11. How much faster is a Pentium microprocessor than an 8086?
12. How is information stored on a floppy disk?
13. What are the input/output ports on a computer used for?
14. Which type of monitor is most prevalent on today's computers?
15. Describe the four types of printers in terms of print quality.
16. Which type of printer can print in multiple colors?
17. Briefly explain the three methods of obtaining software.
18. What are the six major functions of software?
19. Describe the steps in the process of buying a computer.
20. Suppose you were running a local farm supply business. What type of computer would you buy? Justify your decision.

Chapter 2

OPERATING SYSTEMS

Most computer users find a certain system they like and use it exclusively. They have determined what they want the computer to do and which applications they need to accomplish this work.

The operating system is the software that connects applications with the computer hardware. It is sometimes called the platform. Some computer users are not even aware of the operating system, while others make extensive use of the system to customize their computer for maximum productivity.

In the past, the operating system determined which software a computer user was able to use. Now, most computer software companies develop versions of their software that are compatible with most of the existing platforms. As an example, the WordPerfect corporation has different versions of its popular word processing program, WordPerfect, for computers using MS-DOS, Macintosh System, and Unix operating systems.

This chapter takes a look at the most common operating systems used on today's computers. A description of each operating system is given, and a more extensive description is provided for MS-DOS, the most popular operating system currently in use. Several utilities available that enhance the operating system and hardware are also discussed.

OBJECTIVES

1. Describe the functions of an operating system.

2. Discuss the various means of classifying operating systems.

3. Identify the most popular operating systems and user interfaces.

4. Describe the process a computer undergoes when it is started up.

5. Utilize MS-DOS files.

6. Associate MS-DOS commands with their functions.

7. Associate MS-DOS error messages with their causes and possible solutions.
8. Identify popular utilities programs and their functions.

FUNCTIONS OF OPERATING SYSTEMS

The operating system is a set of files that serves as the connection between the user, application software, and the computer hardware. It coordinates the functions of the microprocessor, memory, input/output devices, and the software running on the computer. It does this by providing a standard, or platform, for software developers to use when writing programs.

The central core of information, called a *kernel*, of the operating system must reside in RAM at all times. This is the information that manages the computer for the applications running under the operating system.

The operating system also contains various commands that the user can use to manage files and disks, optimize memory, or configure hardware. Knowledge of these commands and their functions makes a user proficient in using an operating system.

The three primary functions of the operating system are allocation of

TERMS

address	filename	prompt
AUTOEXEC.BAT	IBM compatible computer	pull-down menu
booted		root directory
buffer	internal commands	spooling
command files	kernel	subdirectory
CONFIG.SYS	login	switch
conventional memory	logout	system security
device drivers	Microsoft Windows	time slices
extension	multitasking	user-friendly
external commands	on-line time	virtual memory management
file allocation table (FAT)	parameter	
file	point-and-click	virus protection
file management	program file	wildcard

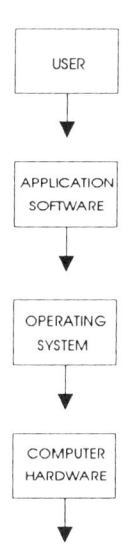

Figure 18-1. The operating system keeps the user and application connected to the computer hardware.

computer resources, monitoring of system activities, and management of files.

ALLOCATION OF COMPUTER RESOURCES

The operating system manages the computer's microprocessor, memory, input/output, and storage access to keep the programs and instructions from getting mixed up during processing.

Microprocessor

The microprocessor can only carry out one instruction at a time, although it does so very quickly. When the microprocessor receives more than one instruction, the operating system coordinates the activity. This happens frequently when more than one user or application is trying to access the same processor, although it can happen to a lesser extent with one user and one application.

The operating system allocates *time slices* to each user and application using the microprocessor. These time slices are measured in milliseconds. Each user typically receives a time slice in rotating order. In newer machines with fast microprocessors, the users don't realize that they are sharing with someone else because in real time the microprocessor stills performs very fast.

Memory

A computer's RAM must hold several things at one time. The operating system, the instructions from application programs in use by one or more users, data being processed, data being sent to output devices, data being received from input devices, space for procedures such as calculations and sorting, and TSR programs all must be available in RAM. The operating system keeps this information sorted. The system's primary function is to make sure that information from one application or user does not replace other information in memory that is needed by another application or user.

The operating system may do this in one of two ways. One is to maintain an *address* of all information located in RAM at any one time so that the operating system can determine RAM space that is available. The other way is to use *virtual memory management*. In virtual memory management, the operating system increases available memory by allocating some of it to a hard disk. This method is commonly used in systems with multiple users.

Input/Output

While input devices such as the keyboard, mouse, modem, and disk drives may be sending information to the computer, output devices such as the monitor, printer, mouse, and disk drives may be receiving information from the computer. The operating system coordinates these processes.

The operating system handles this information by creating *buffers* and by *spooling*. The buffers are parts of RAM used for temporary storage of information. They are usually used for information going to disk drives and monitors, while spooling is used for information going to slower devices, such as printers and modems. When spooling, the operating system sends some of the information to temporary storage on the hard disk (or a floppy disk).

Storage Access

The operating system also gives a user or users access to storage devices such as disk drives and tape drives. Disk drive access is usually given on a first come, first served basis, with each user getting his or her turn as the instructions are processed. Tape drive storage needs more management because it is a sequential access device. The tape drive is usually reserved for specific users or applications.

MONITORING OF SYSTEM ACTIVITIES

With systems that have more than one user, the operating system may be required to monitor the time of use and provide system security. Some network systems are set up so the users pay for *on-line time* or for microprocessor time. In these systems, the user must *login* to the system to begin using it and *logout* when finished in order for the system to keep an accurate record of time used.

Having to login and logout leads to some possible security problems. The operating system manages *system security* by requiring passwords for login. The password should be known only to the user. (Of course, the system operator should have the ability to delete passwords when someone quits using the system.) The user should be careful not to let others know his or her password and thereby have access to confidential information or send messages using his or her login name. Some systems are set up with default times so that a user can be timed out when not using the system for a certain period of time. This prevents someone else from coming in and using a person's access because they forgot to logout when finished.

MANAGEMENT OF FILES

The operating system provides the user and application with a set of *file management* techniques that help them keep their data organized. These techniques include copying, saving, renaming, deleting, and sorting files, formatting disks, and moving files from one storage device to another. The applications also have access to these file management functions. For example, when using a word processing program, the user can usually save, delete, rename, move or copy files; in most cases, the application uses the operating system commands to carry out these functions.

Most operating systems also include a file editor for editing certain command files and system files. These file editors can also be used to create files. Specific file management techniques are described later in this chapter.

CLASSIFICATION OF OPERATING SYSTEMS

The different operating systems are designed for various purposes. Computer operators must select a system based on its purpose and the availability of needed applications for the system. Usually, the operating systems are classified in one of three ways. These classifications include the number of tasks or programs the system can run at one time, the number of users the system can handle at one time, and the type of interface provided to the user.

NUMBER OF TASKS

The two categories in this classification are *multitasking* and single tasking systems. Multitasking refers to the operating system's ability to concurrently execute more than one application. For example, on a multitasking system, the user can have a word processing program open and an electronic spreadsheet open at the same time. The user can be printing something with the word processing program and entering data in the spreadsheet at the same time. The microprocessor must switch back and forth between application instructions.

Of the more popular operating systems, Unix, OS/2, and System (Macintosh) are all multitasking systems. MS-DOS is a single tasking system.

NUMBER OF USERS

Computer systems are usually divided into single user systems and multi-user systems. By nature, multi-user systems must also be multitasking systems, where more than one application can be used at the same time. Multi-user systems usually will also allow more than one individual to use the same application. This function is very important to a catalog sales operation, for example, where several operators are taking orders using the same ordering program.

Unix and OS/2 are both multi-user systems. MS-DOS and System are single user systems.

INTERFACES

The operating system's interface refers to what the user sees on the monitor and how the user interacts with the system. A graphical user interface (GUI) refers to an interface characterized by use of a mouse to *"point-and-click"* on icons or to access *pull-down menus* to give commands to the system. A character-based user interface (CUI) refers to a system where the user gives commands by typing them at a *prompt*.

System and OS/2 have GUIs. Unix and MS-DOS are considered to have CUIs, although newer versions of both have incorporated pull-down menus and other GUI features. Microsoft Windows, an operating environment used with MS-DOS computers, is a graphical user interface with multitasking capabilities.

POPULAR OPERATING SYSTEMS AND ENVIRONMENTS

MS-DOS

Microsoft Disk Operating System (MS-DOS) is usually referred to as DOS for short. DOS was developed as PC-DOS by Microsoft for IBM for use on the original IBMpc in 1981. DOS was later marketed as MS-DOS by Microsoft for companies developing *IBM compatible computers* or IBM clones.

DOS quickly became an industry standard in the early 1980s. The leaders at IBM and Microsoft decided to allow anyone to have access to the code, and thereby, any computer programmer could write programs for DOS machines. This led to much more software being written for it than any other operating system, which in turn made DOS-based machines big sellers. The rest of the computer world has yet to catch up. DOS is by far the most popular operating system, running on over 80 percent of all desktop computers in operation. It also has the widest variety of software programs written for it of any operating system.

DOS is actually a 16-bit revision of an 8-bit operating system called CP/M (short for control program/microcomputer). CP/M was developed by Gary Kildall, the founder of Digital Research. Today, Digital Research makes an operating system called DRDOS, which competes with MS-DOS and will run the same software.

The original version of DOS was MS-DOS 1.0. In 1993, Microsoft introduced version 6.0. In between, several versions have been developed. Versions 2.1, 3.3, and 5.0 were very popular. The most commonly used DOS commands are described later in this chapter.

Figure 18-2. Microsoft's version 6.0 of MS-DOS, the world's most popular operating system software.

MICROSOFT WINDOWS

Microsoft Windows, or Windows for short, was the DOS community's answer to the popular GUI introduced by Apple for their Macintosh computer. The GUI is considered more *user-friendly* than the CUI used by DOS for many years. Windows works as an operating environment, providing a multitasking graphical user interface that still uses DOS operating system commands.

Windows became very popular in the early 1990s. Users like it because they can still exit it to run their DOS software, and also take advantage of its graphical capabilities with newer, Windows-versions of programs with which they are already familiar. Windows 3.0 was one of the best-selling programs of 1992 and 1993, and software companies have invested major resources into developing programs for use with Windows. Most popular DOS programs, for example WordPerfect, Lotus 1-2-3, and DBase IV, have Windows versions available.

As an operating environment, Windows acts as integrated software, with common commands for different applications that requires the user to learn fewer commands to run applications. This gives the user the ability to use powerful applications without as much training.

OS/2

In 1988, IBM introduced the IBM PS/2 which included OS/2, a multitasking, multi-user, operating system. Many computer experts tout OS/2 as the best operating system because it will run DOS programs and Windows programs as they were intended. Even so, OS/2 has not sold as well as expected, probably because of the established base of DOS and Windows users who upgrade primarily with newer versions of those programs.

OS/2 is a very powerful operating system. It is the only popular operating system that combines a GUI (its Presentation Manager), multitasking, and multi-user capabilities. (Windows NT, expected to be released in late 1993 or early 1994, is supposed to do all three as well.) Although it is provided with new IBM computers, IBM's share of the PC market is not as extensive as it was in the 1980s. Within the next few years, however, all PC's will probably be sold with an operating system like OS/2, although it may be simply an upgrade of DOS.

SYSTEM (Macintosh)

In 1983, Apple introduced the Lisa computer, an innovate product that never sold well because of its $10,000 price. Lisa was discontinued in 1984. That same year Apple took the best parts of the Lisa and packaged them into a lower-priced computer system called the Macintosh (Mac for short). The Mac was also overpriced and scoffed at by IBM and compatible users as a toy, fun to play with but weak in computing power.

One thing the Mac had going for it, however, was the first GUI, which was easier to understand and use than DOS's command line. For the first time, novice computer-users did not have to know what command to type and how to type it. They could simply select from a menu of commands or choose an icon by clicking on it with the mouse.

At first, Mac users were almost like a cult, refusing to learn DOS commands in order to use a computer, but without many software options available to them. At first, Apple wrote its own programs for the Mac, and other software companies were not invited. Then, even after Apple paved the way for other companies to write software, the companies did not invest much time and effort in Mac applications. DOS had a much greater base of users, and good DOS-based applications were sure to sell more than those developed for Macs.

The Macintosh overcame these early hurdles because of agreements with some software companies and a small but growing base of users who were loyal to the Mac way of computing. By the 1990s, Macs had made inroads into business operations that were formerly all-IBM.

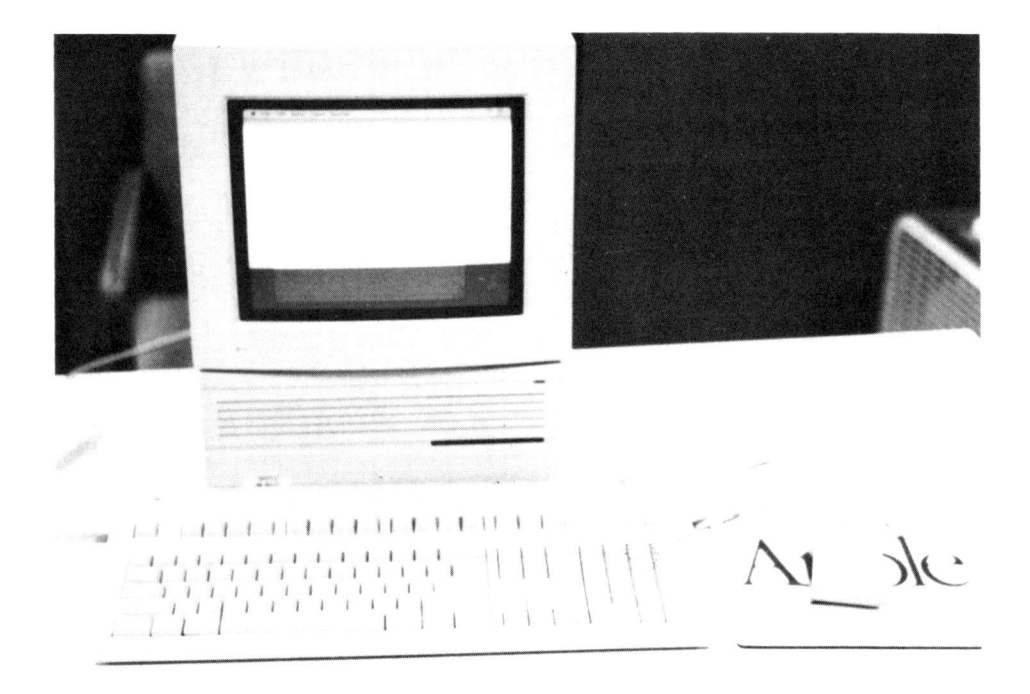

Figure 18-3. The Macintosh from Apple Computers was the first computer to use the graphical user interface (GUI), touted for its ease of use.

UNIX

Unix was developed in the early 1970s by scientists at Bell Laboratories, a part of AT&T. AT&T, under federal regulations, couldn't commercially market Unix until its divestiture as part of the deregulation of the telephone industry in the early 1980s. Prior to this time, Unix was a popular system in use on many university campuses as a multi-user, multitasking system suitable for large networked computers. Many of its features were forerunners of features in the popular operating systems for personal computers.

Unix has grown in popularity since coming on the commercial market in the 1980s. Versions of Unix are run on almost every size of computer, although it is still primarily used in large networks. For this reason, Unix is discussed further in Chapter 23, Communications and Networking.

STARTUP OF THE COMPUTER

A knowledge of what the computer does when started up, or *booted*, will help the user to understand how the operating system works and the importance of various operating system commands and files. A description of the startup of a computer using the MS-DOS operating system is given below. Other systems will be different, but will execute most of the same functions.

LOADING THE SYSTEM

When the power to the computer is switched on, it first accesses ROM memory to give the initial instructions to be followed. These initial instructions are called the boot record. The boot record tells the computer to go to the default disk drive and load the operating system into the computer. On most computers the default drive is the A: drive, usually a floppy disk drive. If the A: drive contains a system disk, the system is copied from the disk to the computer's RAM. If not, the computer will usually go to the hard disk, drive C:, and get the system. This procedure is important in that the user will normally have the system wanted on the hard disk, but if he or she wants to change for some reason, they can use the A: drive to boot with the different system.

The boot procedure loads three DOS files into RAM. The three files are COMMAND.COM, IO.SYS, and MSDOS.SYS. Having these three files on a disk makes it a system or startup disk. A directory listing of the files on the disk will only show the COMMAND.COM file because IO.SYS and MSDOS.SYS are hidden files. These files contain the kernel of the operating system that resides in RAM at all times.

CUSTOMIZING THE STARTUP PROCESS

After loading the system, the startup procedure can stop after loading the system at an A: or C: prompt. At this point the user enters a command or executes a program to continue use of the computer. Most users, however, have utilized advanced features of DOS to customize their startup procedures. Two files give users this flexibility: CONFIG.SYS and AUTOEXEC.BAT. When the computer has the system loaded, the boot procedure tells it to look for these files and execute them as part of the startup process.

The *CONFIG.SYS* file is created when DOS is installed to the computer, but the user can edit this file to customize his or her system. CONFIG.SYS can be customized to do many different things. Some of the more common ones include: (1) indicate a country code, with symbols and conventions pertaining to a particular country; (2) specify different *device drivers* to handle memory or hardware; (3) specify the number of files that may be open at one time; (4) load DOS and various TSRs in high memory; (5) specify the number of buffers in RAM; and (6) indicate the location of files and other information.

Figure 18-4. To interact with peripherals such as these tape drives, the operating system must have device drivers, special programs written for each type of peripheral.

The *AUTOEXEC.BAT* program is created by the user of a system to save time during the startup process. AUTOEXEC.BAT contains a list of commands which are executed at startup, rather than the user having to execute each command individually every time the system is booted. These commonly include commands to establish a current path and/or default directory, set screen attributes, start TSR programs such as screen savers, run disk utilities, customize the prompt, and access a menu or operating environment.

MS-DOS FILES

A *file* is a collection of logically related information stored together on a disk. The information is stored magnetically, much like a tape recording. The location of a file (which sectors and tracks contain the file) on a disk is called the address. When a file is written to a disk, the address is recorded in the *file allocation table (FAT)*.

Files may be either data files, program files, or command files. Data files contain information entered by the user of the computer system. *Program files* are files used to operate an application program. *Command files* are files used by the operating system and utilities.

DOS FILENAMES

DOS files may have a *filename* and *extension*. The filename can be up to eight characters in length, the extension up to three. The filename and extension are separated by a period. Extensions are optional; they are recommended for naming files and not recommended for naming directories. Filenames may be typed in upper or lower case letters. MS-DOS converts the characters to upper case.

Valid characters for filenames include all letters and numbers and these symbols:

$ % ' - @ { } ~ ' ! # ( ) &

Spaces cannot be used in filenames. Also, several symbols cannot be used. The following symbols *cannot* be used in naming files:

. " / \ [ ] : | < > + = ; ,

These symbols are used for other purposes in DOS and might confuse the system if used in filenames.

When using DOS commands, a *wildcard* may be used. Wildcards include the asterisk (*) and the question mark. The asterisk is used to represent multiple letters in filenames and extensions. The question mark is used to represent a single letter in a filename or extension. For example, *.* is used to include all files, while *.txt will include only filenames with the extension .txt. On the other hand, ?.* will include only one-letter filenames, while *.t?? will include all files with an extension beginning with a "t."

DOS FILES EXECUTION

Some DOS files can be executed by typing the name of the file and pressing <ENTER>. These files, called executable files, can be identified by their extension. Executable files will have one of three extensions: .COM, .EXE, or .BAT. When naming data files, the user should avoid using any of these extensions.

.COM Files

The .COM extension is usually reserved for external DOS commands and utilities command programs. For example, the DOS command to format a floppy disk is actually an executable file named FORMAT.COM. Typing FORMAT at the prompt will execute the file and begin the process of formatting a disk.

.EXE Files

The .EXE extension is used primarily by applications programs to indicate files that will start up (boot) some portion or all of the program. For example, the file that starts WordPerfect is WP.EXE. Typing WP at the prompt will execute this file and begin WordPerfect.

.BAT Files

Files with the .BAT extension are usually created by the user and are called batch files. Batch files are a collection of DOS commands and procedures stored in a file. They help the user take short cuts and save time. Instead of typing in each command, batch files allow the user to type a single command which tells DOS to run the entire batch of commands. The AUTOEXEC.BAT file mentioned earlier is an example of a special batch file that is executed automatically by DOS whenever the system is booted.

The computer user can create batch files to perform many tasks. Batch files are used to provide the menus and options for different situations. If a user has to enter multiple commands in sequence on a regular basis, he or she has a use for a batch file. This is often true when loading a special program or looking for a menu of possible executable files on a disk.

Although almost any DOS command can be included in a batch file,

OPERATING SYSTEMS

some special commands make batch files more efficient for the user. A brief description of each follows.

CALL—executes another batch file, then continues with the current batch file.

ECHO—displays messages on the screen. ECHO OFF hides subsequent commands and messages.

FOR—executes a series of commands.

GOTO—jumps to the designate label.

IF—a conditional statement, based on input from the user.

PAUSE—temporarily stops the execution of the batch file.

REM—used to imbed a remark or comment into the batch file. The user cannot see the comment during normal execution of the batch file.

SHIFT—moves the command line arguments one position to the left.

```
ECHO OFF
CLS
PATH=C:\;C:\DOS;C:\WP51;C:\UTIL;C:\NAV;C:\DBMENU;
MOUSE
PROMPT $P$G
MENU.BAT
```

Figure 18-5. An example of an AUTOEXEC.BAT file, read as part of the startup procedure.

MS-DOS DIRECTORY STRUCTURE

In DOS, all disks have a directory, much like a table of contents, where the information is listed in the FAT. In many cases, however, having one directory is not a good way to keep files organized. In addition to having the *root directory* (created when the disk is formatted), the user may also create a *subdirectory* for each group of files that contain related information.

For example, the WordPerfect word processing base program contains over 100 files. Rather than putting all of those files in the root directory, most users will create a subdirectory just for WordPerfect files. Under that directory, the user will probably create more subdirectories for graphics files and document files. The document subdirectory may be further divided into more subdirectories for individual projects and the files that relate to them.

Another way to designate the relationship is to refer to directories as parent and child directories, which relates to a family tree. If a directory is created within another, the created directory is referred to as a child directory of the original directory, while the original directory is a parent directory. Just like a family tree, a child directory must have a parent directory. In the example given above, the WordPerfect directory would be a child directory of the root directory, but a parent of the WordPerfect documents directory.

Each directory has a path, a way to determine the tree leading to that particular directory. Some common paths are given below.

C:\—The root directory

C:\WP51—WordPerfect subdirectory (child of the root directory)

C:\WP51\DOCUMENT—Document subdirectory (child of the WordPerfect subdirectory)

C:\WP51\DOCUMENT\LETTERS—Letters subdirectory (child of the Document subdirectory)

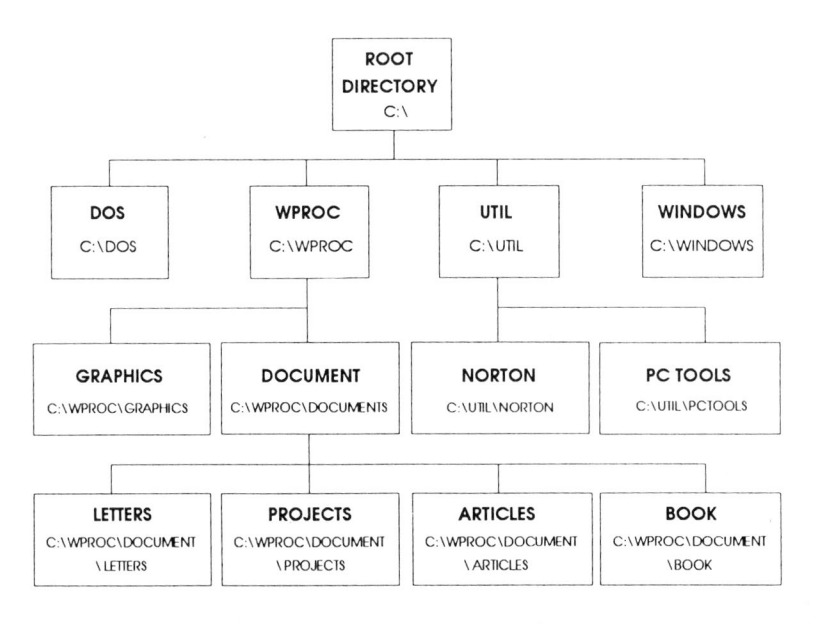

Figure 18-6. A possible directory structure (tree) for a hard disk designated as drive C:. The path for each directory is given under the directory name.

MS-DOS COMMANDS

Common DOS commands allow the user and applications software to participate in file, environment, and hardware management. These commands may be typed in at the prompt or accessed through the DOS shell (or Windows).

DOS commands take a general form: the command, a space, the *parameter*(s) (if any), a space, the *switch*(es), (also called options, if any), and press <ENTER> to activate the command. As an example, the DOS command for listing the files is DIR. To list files in the current directory, the user would enter:

DIR

The user often is working in one directory and wants to see a list of files in another drive or directory. The other drive is listed as a parameter of the DIR command. If the user is working in A: drive but wants to list files in the DOS directory of drive C:, he or she would enter:

DIR C:\DOS

When a directory contains more files than can be listed on the screen at one time, the above commands will allow the first files listed to scroll off of the screen without giving the user time to read the filenames. This problem can be remedied by adding a switch or option to the command. To have the listing of files stop when the screen is full and wait for the user to go to the next screen, the user would enter:

DIR C:\DOS /P

Note that the switches are separated by a slash (/). The space is optional.

Unlike some other operating systems, DOS does not distinguish between upper and lower case letters when processing commands. DOS actually changes lower case commands to upper case before processing them.

DOS has two types of commands: internal and external. Some of the most commonly used commands are described below. The user should consult a DOS manual for more details about these and other commands.

INTERNAL COMMANDS

The commands most commonly used by users and application programs are called *internal commands*. These commands are loaded in RAM when the system is booted and are part of the DOS kernel that must be available in RAM at all times. These commands and their functions are described below.

COPY

This command duplicates a file from one disk or directory to another or from the keyboard to a disk. It requires parameters of the source directory/filename and target directory/filename. (The current directory is the default.) No switches are used with COPY. To copy a file named CHAPTER.TXT from the drive A: root directory to the DATA directory in drive B:, the user would enter:

COPY A:\CHAPTER.TXT B:\DATA\CHAPTER.TXT

The target filename can be changed if desired. If left off entirely, the target filename will be the same as the source filename.

With the COPY command, the wildcards * and ? can be used.

CHDIR (CD)

The CD command changes the current directory to the directory specified. For example, the command

CD C:\DOS

will change the current directory to the DOS subdirectory in drive C:. The CD command requires a parameter of the desired directory but the full path is optional if going up or back one level in the current directory structure. Some other parameters include .. and \. As examples,

CD ..

changes the current directory back one level to the parent directory, and

CD\

changes the current directory to the root directory in that drive.

CLS

The CLS command clears the screen. Although it can be used at the prompt, it is used almost exclusively in batch files to clear the screen after a command has been processed. The user generally inserts the batch file command PAUSE before using CLS, because CLS will clear the screen before the results of the previous command can be seen.

DATE

The DATE command allows the user to change the date in the computer. In computers with battery backups, this command is seldom needed, except when the battery runs down. In computers without a battery backup, the DATE command is often part of the AUTOEXEC.BAT file, allowing the user to input the date each time the system is booted.

DEL (ERASE)

The DEL and ERASE commands are interchangeable. Both remove a file or files from a directory. The parameter required is the directory and filename. The current directory is the default directory, there is no default filename.

The wildcards * and ? can be used to remove multiple files, but should be used carefully. For example,

    DEL *.*

will remove all of the files from a directory.

DIR

The DIR command is a very useful command in file management. It provides a list of the files in a directory so the user can determine which files to move, copy, delete, etc. The parameter required is the directory; the default is the current directory. Switches include /P to keep the list from scrolling off the screen, /W to give filenames in more than one column, /A to select by attributes, or /O to select in some specified order. For example,

    DIR C:\DOS /W

will provide a list of the files in the DOS directory in drive C: in a wide column format.

MKDIR (MD)

The MD command creates a child directory (also called subdirectory) of the current directory. This allows the user to keep files that are related in a specific place for easy access. The command parameter is the name of the directory. Directories can be named just like files, but most experienced DOS users have adopted a practice of naming directories without extensions and files with extensions. This practice makes it easier to tell at a glance whether a name is referring to a file or a directory.

PATH

The PATH command tells the computer what directories, other than the current directory, should be searched when looking for a filename. This command is commonly used in AUTOEXEC.BAT files to give a list of probable directories. A common PATH given in an AUTOEXEC.BAT file might look like the following:

    PATH C:\;C:\DOS;C:\UTIL;D:\WPROC

Loading this path in when the computer starts tells it to look in each of these directories when searching for a command or file not found in the current directory.

PROMPT

The PROMPT command changes the prompt seen by the user. It is commonly used in AUTOEXEC.BAT files. The default prompt is A or C. If the user wants the prompt to display the current drive and directory and then keep the greater-than sign at the end, this line should be part of the AUTOEXEC.BAT file:

    PROMPT PG

RENAME

The RENAME command changes the name of a file. The parameters are

OPERATING SYSTEMS

the old filename's drive and directory and the new filename's drive and directory. The default is the current drive, with no default filename. This command can be used to move files as well. For example, to move a file called CHAPTER.TXT from the root directory in drive A: to the root directory in drive B:, the user would enter

    RENAME A:\CHAPTER.TXT B:\CHAPTER.TXT

RMDIR (RD)

The RD command removes the specified directory, if it is empty (containing no files or subdirectories). The current directory must be the parent directory of the one that is being removed, or the entire path for the directory to be removed may be entered.

TIME

The TIME command is similar to the DATE command in that is it seldom needed in a computer system with a battery backup and used as part of the AUTOEXEC.BAT file in systems without a battery. The TIME command is used more in systems with a battery than DATE, however, because the internal clock in computers loses a little time each day. The user will find it necessary to adjust the clock periodically using the TIME command.

TYPE

The TYPE command prints the specified file on the screen. The parameter is the location and name of the file. The type command can also be used to print the file to a printer or other output device or to a disk using the > (redirect) sign. For example, the command

    TYPE A:\CHAPTER.TXT

will print the contents of the file CHAPTER.TXT on the screen. To send the file's contents to the printer, the user would enter

    TYPE A:\CHAPTER.TXTPRN

(The redirect command [>] can also be used to send output from other commands to the printer or to a disk file.)

If the file is a long file that scrolls off the screen, the MORE command can be used in conjunction with the type command to stop the scrolling at the end of each full screen. To use this to view the file CHAPTER.TXT on the screen, the user would enter

TYPE A:\CHAPTER.TXT|MORE

Note that the symbol used is the pipe (|) symbol, not a colon (:).

Table 18-1
QUICK SUMMARY OF INTERNAL MS-DOS COMMANDS

Command	Function
COPY	Makes a duplicate copy of a file
CHDIR (CD)	Changes the current (working) directory
CLS	Clears the screen
DATE	Allows user to set the date
DEL (ERASE)	Removes a file from a disk
DIR	Lists the files in a directory
MKDIR (MD)	Makes a new directory
PATH	Changes the current path
PROMPT	Allows user to customize prompt
RENAME	Changes a filename
RMDIR (RD)	Removes a directory from a disk
TIME	Allows user to set current time
TYPE	Sends contents of a file to screen

EXTERNAL COMMANDS

External commands are not used quite as commonly as internal commands and are not part of the operating system kernel. Each external command has its own separate filename as part of the DOS files. To use these commands, the files for the commands must be in the current directory or the directory containing the commands must be in the system's current path (as described in the section on the PATH command) so the system will know where to look for the commands.

Some of the more commonly used commands are described below. The user should consult a DOS manual for more information about specific commands.

BACKUP

The BACKUP command makes a copy of specified files or directories, compressing them so fewer disks are used than with the COPY command. This procedure is essential to hard disk users, who should *always* have a backup of the information on their hard disk. Hard disks (floppies, too) sometimes crash and lose all of their data.

Older versions of DOS had a poorly-developed BACKUP function and experienced users usually had a utility for this process. Version 6.0 has a very usable backup system designed like the ones used by the utilities. It has pull-down menus and allows the user to specify files or directories to be backed up.

CHKDSK

The CHKDSK command performs routine checks of the status of a disk. It provides the user with a summary of the total space on the disk, the space available, what the space is being used for, total memory of the system, and available memory. It will also give the serial number of the disk.

DBLSPACE

DBLSPACE is a disk compression utility that is new with DOS 6.0. It allows DOS to compress files before writing them to a disk and uncompress them after reading them from a disk. With a compression ratio of approximately 1.8:1, this utility allows the user to store almost twice as much information on a given disk.

DEFRAG

As files are written to and deleted from a hard disk, it becomes fragmented. This means that files may be located in more than one sector on different parts of a disk. After a file is deleted, the space it was using is open to the next file saved to the disk. The new file may be longer (or shorter) than the deleted file. DOS will write some of the file in the space where the deleted file was located, and then write the rest of it to a new location. As more and more files are written this way, the disk operates slower than if each file was written to one location. The DEFRAG utility rewrites the entire disk with files all moved to one location. This speeds up access time when using the hard disk.

DISKCOPY

The DISKCOPY command copies the entire contents of a disk to another identical size and type disk. The command erases all of the information on the disk being copied to, if it has any. DISKCOPY will also format the disk being copied to, if it is not already formatted. If a computer has two of the same type disk drives, the command can be entered as

DISKCOPY A: B:

If the computer has only one disk drive of each type (as is most common on newer computers) then the user can enter

DISKCOPY A: A:

DOSSHELL

The shell program in DOS is its answer to the popular GUI environments many users prefer. A mouse and pull-down menus are used to execute commands and programs. The shell contains a directory tree of the current directory, a list of files in the current directory, a menu bar with pull-down menus, icons for changing to the various disk drives in the system, and a selection of utilities for the user. DOSSHELL is actually the name of the .EXE file that executes when the command is entered by the user.

Most users who use the shell include the DOSSHELL command as the last line of their AUTOEXEC.BAT file. This tells the computer to enter the shell after completing the startup procedure.

EDIT

In versions 5.0 and 6.0, DOS includes a text editor with pull-down menus for editing ASCII files, such as batch files, from the DOS prompt. The EDIT command activates this editor.

FDISK

The FDISK command is used to establish partitions in a hard disk. It is used when setting up a new hard disk or when changing the partitions in an existing hard disk. Either way, any information saved on the hard disk will be erased when running FDISK.

FORMAT

The FORMAT command prepares a disk for use by dividing it into sectors and tracks. It will erase any information that might be on the disk. The parameter is the drive. Switches include /S to make the formatted disk a system disk (with the DOS kernel on it), /V to prompt the user for a volume label on the disk, /Q to quickly format an already-formatted disk by erasing the file allocation table, and /B to save room for the system without making it a system disk. For example, to format a disk in drive A: and make it a system disk with a volume label, the user would enter:

    FORMAT A: /S /V

HELP

The HELP command is used when the computer user needs quick information about other DOS commands. The parameter is the command about which the user wants information. For example, when information about the TYPE command is needed, the user would enter

    HELP TYPE

and DOS would give the user a brief description of the command. The HELP command is not designed to replace the DOS manual, however.

MEMMAKER

Older versions of DOS recognized 640K as the maximum memory available in RAM. The first 640K is referred to as *conventional memory*. As programs started requiring more memory, a way to address more memory using DOS was needed. After conventional memory, DOS recognizes the next 384K as expanded memory. Any memory beyond that is referred to as extended memory.

The MEMMAKER program analyzes the memory available in a system and allocates DOS and other programs to maximize the memory available to the user. This procedure is important because many applications are designed to run in conventional memory only.

MORE

The MORE command is a filter command often used with the TYPE command to keep long files from scrolling off the screen. In newer versions of DOS, however, the MORE command can be used by itself to view large files. To view a file called CHAPTER.TXT, 24 lines at a time, the user would enter

>   MORE<CHAPTER.TXT

and the file would show up on the screen with "—More—" shown on the last line. To view the next screen of information, the user would strike any key.

MSAV

The MSAV command activates the *virus protection* utility available in DOS 6.0, allowing the user to scan disks for viruses. The parameter for the command is the disk drive. DOS 6.0 also has a TSR program—VSAFE.COM—that a user can load using the AUTOEXEC.BAT file which will provide ongoing protection from the introduction of viruses through floppy disks.

PRINT

The PRINT command allows the user to make a printed copy of an ASCII text file. To print a copy of the AUTOEXEC.BAT file, for example, the user would enter

>   PRINT C:\AUTOEXEC.BAT

and the PRINT command will send the specified file to the printer. Of course, the same result can be obtained by using the TYPE command and redirecting it to the printer. (TYPE C:\AUTOEXEC.BATPRN)

RESTORE

In versions of DOS prior to 6.0, the RESTORE command is used to replace files copied using the BACKUP command. The parameters are the drive containing the disks the information was backed up to and the destination of the files to be restored. To restore files backed up to drive B: from C:\WPROC, the user would enter

    RESTORE B: C:\WPROC

and the system would copy the backed up files from drive B: to the C:\WPROC directory. DOS will replace any files with the same name on C:\WPROC with the backed up files, so users must be careful.

In DOS 6.0, the restore function has become more integrated with the BACKUP function, complete with pull-down menus for selecting drives and directories.

SYS

The SYS command copies the hidden system files (IO.SYS & MSDOS.SYS) to the specified disk. The parameter is the disk drive specification. For example, to copy the system files to a disk in drive A:, the user would enter

    SYS A:

The SYS command does not copy COMMAND.COM, the other file necessary to make a disk a true system disk.

TREE

The TREE command lists all of the directories and subdirectories on a disk. The parameter is the disk drive specification. TREE can also be used to list the files within the subdirectories by using the /F switch. To list all the files in all of the directories and subdirectories on drive A:, for example, the user would enter

    TREE A: /F

UNDELETE

The UNDELETE command allows users to rectify a common error: erasing an important file by mistake. The parameter is the directory from which the file was erased. Undeleting a file is possible because the DEL or ERASE command does not actually erase the file from a disk, just its address from the file allocation table (FAT). As long as no other file has been written over it, a deleted file remains on the disk. When the user enters an UNDELETE command, such as

UNDELETE A:\

DOS will list the deleted files that can be recovered. It will prompt the user about whether or not to restore each deleted file.

Table 18-2
QUICK SUMMARY OF EXTERNAL MS-DOS COMMANDS

Command	Function
BACKUP	Makes backup copies of files/directories
CHKDSK	Checks disk space and system memory
DBLSPACE	Compresses information on a disk
DEFRAG	Rewrites disk files to defragment
DISKCOPY	Makes a duplicate of a disk
DOSSHELL	Allows user to use the shell interface
EDIT	Allows user to edit an ASCII file
FDISK	Partitions a hard disk
FORMAT	Sets up a disk to receive information
HELP	Provides information about DOS commands
MEMMAKER	Optimizes available memory of a system
MORE	Stops output at 24 lines (one screenful)
MSAV	Scans a disk to detect viruses
PRINT	Prints the contents of a file
RESTORE	Restores files backed up with DOS 5.0 or earlier
SYS	Loads the system files on a disk
TREE	Provides a list of directories and subdirectories
UNDELETE	Restores a file that has been removed from a disk

ERROR MESSAGES

DOS has two versions of error messages. When the shell is being used, a command error will result in the error window being shown with the type of error and options for the user, usually including: try again, don't try again, OK, Cancel, and Help. When an error is made in a command line entry, the type of error is shown on the screen and options may or may not be given, depending on the message.

Some of the common error messages and prompts are given below, with the reason for the message and possible solutions to the problem.

Abort, Retry, Fail?

Abort, Ignore, Retry, Fail?

These prompts follow a number of error messages, including disk drive not ready, and printer not found. A prompt selection can be chosen by typing the first letter. The ignore and fail options are used mainly by computer programmers.

Problem: DOS can't read a disk or communicate with a peripheral, such as a printer.

Solution: First, try to fix the problem, such as closing the door on the disk drive. Then type R several times. Then try A. Most users should avoid I or F as this can result in data loss.

Bad command or filename

Problem: DOS can't recognize the command or filename typed. This is the most common error message. Misspellings, forgetting to use the DOS command before the filename, or being in the wrong directory are common reasons.

Solution: Check spelling of command or file, do a DIR to check which directory is current and the location of the desired file.

Attempted write protect violation

Problem: Disk is locked by the write/protect tab or switch.

Solution: Remove write-protect tab from 5.25-inch disk, move write-protect switch on 3.5-inch disk.

File not found

Problem: The DOS command worked, but the file name specified is not in the directory specified.

Solution: Check filename spelling, do a DIR to see if file is where expected.

Invalid directory

    Problem: DOS can't find directory named in the command.

    Solution: Check spelling of directory name, do a DIR or TREE to see if directory exists or if the right path is selected.

Non-system disk or disk error
(Replace and strike any key when ready)

    Problem: The disk being accessed does not have the system on it. Usually occurs because a data disk was left in the A: drive when booting the computer.

    Solution: Reboot with a good system disk or remove data disk from drive A: so the computer will go to C: where the system is located.

UTILITIES FOR MS-DOS COMPUTERS

Over the years, experts have found areas in which they needed a little more than DOS was providing. Several software vendors have met these needs with programs called utilities. Utilities are designed to allow users to optimize their existing hardware and software by making it more powerful and easier to use. In turn, DOS has included some of these capabilities in its newer versions and will continue to do so. In many cases, however, the DOS version does not have all of the features available in some of the specialized programs from other vendors.

Utilities programs may be very simple or very complex. Their functions may be very specific or general. Some utilities programs fall into the expensive category of commercial software. Others may be user-supported software or freeware.

Common functions of utilities software include file management, hardware diagnostics/optimization, virus protection, memory management, and menu building. These functions are described below. Each description includes the appropriate DOS utility name (if any) and the name of popular programs by other vendors.

FILE MANAGEMENT

File management refers to the ability to copy, view, delete, move, and rename files. Also included is the ability to undelete files. Many utilities offer file management. Most don't do anything special that can't be done using DOS, they just offer the user another way to do these things that the user may find a little easier. DOS has developed its shell in later versions to perform basically the same function.

Two of the most popular utilities programs are Symantec's Norton Utilities, designed by computer guru Peter Norton, and Central Point Software's PC Tools. Both are general programs with a variety of functions and excellent file management capabilities.

HARDWARE DIAGNOSTICS/ OPTIMIZATION

Hardware diagnostics/optimization software includes programs that analyze disk drives to check for defects, such as Norton Utilities, with its Disk Doctor program. Disk Doctor has more features and is much easier to use than the DOS programs used for disk diagnostics and repair. Some users run Disk Doctor as part of the startup procedure each time the system is booted.

This category also includes utilities software that is used for disk data

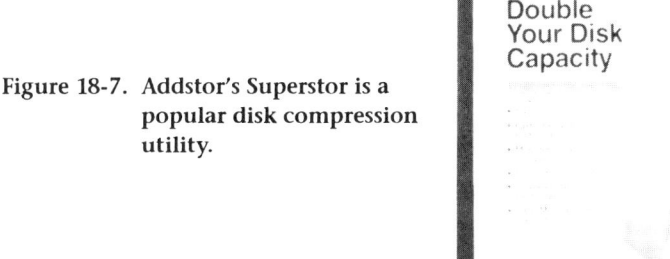

Figure 18-7. Addstor's Superstor is a popular disk compression utility.

compression, such as Stac Electronic's Stacker and Addstor's Superstor PRO. These programs led to DOS 6.0's DBLSPACE utility. Also included in this category would be disk defragmentation utilities such as Norton's SpeedDisk and PC Tools' Compress, another feature DOS 6.0 has included with its DEFRAG utility.

VIRUS PROTECTION

Computer viruses are prevalent in systems around the world. Virus protection is important to avoid loss of programs and data. With version 6.0, DOS incorporated this function with its MSAV and VSAFE utilities. Other excellent virus protection can be obtained from the commercial software Norton Antivirus, Central Point Anti-Virus for DOS, and VIREX, and the user-supported F-PROT.

MEMORY MANAGEMENT

Getting the most out of available memory has been important to users

Figure 18-8. WordPerfect Corporation's Library program provides users with the ability to easily set up and edit menus.

for many years, especially as programs became larger and DOS still would only recognize 640K as conventional memory. DOS 6.0 includes the MEMMAKER function to optimize available memory. Other programs that do this well include Qualita's 386MAX, Quarterdeck Office System's QEMM-386, and, for networked systems, Helix Software's NetRoom.

MENU BUILDING

Menu building programs allow the user to customize his or her working environment by developing menus from which to select applications. Most of the large applications software developers have menu programs available, including WordPerfect, Lotus Development, Microsoft, and many others. Also, many functional programs can be found from shareware distributors and from computer bulletin boards as freeware. Experienced DOS users often develop their own menus using ASCII text files and batch files.

CHAPTER QUESTIONS

1. What are the five most popular operating systems/operating environments in use today? Which one is used by the most people on their personal computers?

2. Briefly describe the three functions of an operating system?

3. What are the three ways operating systems can be classified? Briefly describe each.

4. What are the three system files in MS-DOS?

5. What is the purpose of the CONFIG.SYS file?

6. How does the AUTOEXEC.BAT file save time for the computer user?

7. What is a file?

8. How many letters long may an MS-DOS filename be?

9. What symbols can *not* be used in filenames?

10. What tells the user if a file is an executable file?

11. Write a short batch file that will: (a) list the files in drive A:, (b) clear the screen (after letting the user see the file list), (c) show the contents of a file in drive A: called MENU.TXT, and (d) put in a comment with the filename and purpose that the user cannot see when executing the file.

12. Describe four basic utilities functions. Include the DOS command (if any) for activating these utilities and other programs that have these capabilities.

13. Using the descriptions in the chapter, give the result of each of the following commands:

   a. DIR A:\DATA /W
   b. TYPE C:\WPROC\DATA\CHAPTER.TXTA:\C1.TXT
   c. TREE A: /F
   d. MORE > C:\WPROC\DATA\CHAPTER.TXT
   e. CD..
   f. PROMPT
   g. MD BULLDOG
   h. DISKCOPY B: B:
   i. FORMAT A: /S

Chapter 3

WORD PROCESSING

Word processing is far and away the most common use of personal computers. Over 80 percent of personal computers are used at least partly for word processing. Most offices have one or more computers dedicated exclusively to word processing. Even people who use other applications often spend most of their time word processing.

For several reasons, using a word processing application is more efficient than using a typewriter. With a computer, written documents can be saved and reloaded later for revisions. Similar letters sent to several different clients do not have to be retyped entirely. The user can take advantage of the computer's speed in checking spelling and grammar, and also performing other functions. The computer's flexibility, combined with that of the printer, allows the user to customize letters with graphics and different text sizes and functions.

This chapter describes some of the common functions and features of word processing application software. The interaction between the word processing application and computer hardware, especially printers, is also discussed.

OBJECTIVES

1. Define word processing.

2. Describe the major functions of word processors.

3. Describe the common features of word processors.

4. Describe selected advanced features of word processors.

5. Describe a common page of paper for use in a word processing program.

6. Discuss terms associated with printer typefaces, typestyles and fonts.

WORD PROCESSING AS A COMPUTER APPLICATION

Word processing is the automated processing or manipulation of words using a specialized application program designed to compose, revise, print, and file written documents. It is a specialized form of data processing where the data are characters and words.

In word processing, characters are typed on a keyboard or input into the computer through other hardware. The characters can be grouped into words, sentences, paragraphs, and pages. The largest grouping is called a *document*. Each document generally is stored as a file, although very large documents may be split into several files.

WORD PROCESSING SOFTWARE

Word processing is accomplished by using a word processing application program. Each software package has particular features and interfaces that make it unique, but most have the standard set of features discussed later in this chapter.

The most commonly used word processing package today is WordPerfect, developed by the WordPerfect Corporation. It is a very powerful program with many capabilities. Other highly-rated and popular programs are

---

TERMS

ASCII files	fixed-pitch font	page description
bitmapped font	font	languages
blocking	font renderer	proportionally-spaced
characters per inch (cpi)	footers	font
code	grammar checker	scalable font
concordance file	headers	scrolling
document	macro	search and replace
document layout	margins	text features
editing	merging documents	thesaurus
		undelete

WORD PROCESSING

Figure 19-1. WordPerfect (left) is the most popular word processing program. Microsoft Word (right) is the second most popular word processing program.

Microsoft Word, Ami Professional, WordStar and MacWrite. Many other programs are available as well.

When choosing a word processing program, the user should look for the number of features in relation to the cost, the stability of the vendor, and the support offered by the vendor. One reason WordPerfect is such a popular program is the 24-hour help line offered by the company to assist users who are having difficulties.

COMPUTER HARDWARE REQUIREMENTS

To run the latest versions of word processing applications, the computer must have a hard disk to store the program files. A minimum of 30MB in storage space is recommended, although more will probably be necessary if the computer is used for other applications.

Most word processors make use of spooling to reduce the amount of RAM necessary to operate the program. A minimum of 640K of RAM will handle the program and most files. However, for those users with large files, 1MB of RAM is recommended. Newer versions will probably have more features and require probably 2MB to 4MB of RAM.

Table 19-1.
POPULAR WORD PROCESSING APPLICATIONS SOFTWARE PACKAGES

Package	MS-DOS	Windows	Macintosh
WordPerfect	Yes	Yes	Yes
Microsoft Word	Yes	Yes	Yes
WordStar	Yes	No	No
MacWrite	No	No	Yes
Ami Professional	No	Yes	No
MultiMate	Yes	No	No
PC Write	Yes	No	No
Leading Edge	Yes	No	No
DisplayWrite	Yes	No	No
Q and A Write	Yes	No	No

Word processing can be done on monochrome monitors, which is a good choice because they are considered easy on the eyes. The computer buyer should be careful, however, to make sure other applications can also use the monitor; many other types of programs require a color monitor.

A printer is a must when using a word processing package. The capabilities of the printer are actually one of the most limiting factors in using the software. For letters and documents to be sent outside of the agribusiness, a minimum quality printer would be a 24-pin dot matrix. A true letter quality printer such as a laser or daisy wheel is recommended. Laser printers offer the most flexibility and best quality, but are more expensive. For word processing of text only, a daisy wheel is sufficient, however, graphics and some text features will not be available. Also, in a business where several people are in the same office, laser or inkjet printers are good choices because they operate much more quietly than impact printers like daisy wheel and dot matrix.

MAJOR FUNCTIONS OF WORD PROCESSORS

Word processing applications can be used for almost anything put down in writing. The five major functions of a word processing application are creating, editing, formatting and printing a document, and also document/file handling.

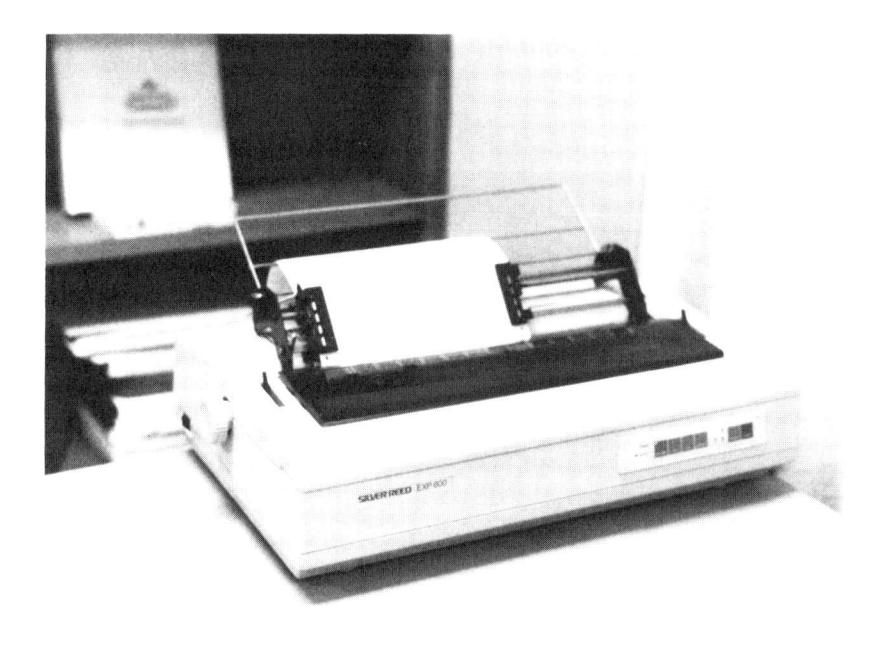

Figure 19-2. A daisy wheel printer gives high-quality text, but will not handle all of the features of a powerful word processor.

CREATING

Creating a document involves opening, naming, and typing or keystroking it. Most word processors start with a blank screen that allows the user to begin typing the new document immediately.

EDITING

Editing a document means making changes to an existing document. This function includes deleting, overwriting, moving, and inserting additional text.

FORMATTING

Formatting a document refers to the word processor's ability to change the look of a document. This includes alignment or positioning of the text.

```
Format
    1 - Line
            Hyphenation                         Line Spacing
            Justification                       Margins Left/Right
            Line Height                         Tab Set
            Line Numbering                      Widow/Orphan Protection

    2 - Page
            Center Page (top to bottom)         Page Numbering
            Force Odd/Even Page                 Paper Size/Type/Labels
            Headers and Footers                 Suppress
            Margins Top/Bottom

    3 - Document
            Display Pitch                       Redline Method
            Initial Codes/Font                  Summary

    4 - Other
            Advance                             Printer Functions
            Conditional End of Page             Underline Spaces/Tabs
            Decimal Characters                  Border Options
            Language                            End Centering/Alignment
            Overstrike
Selection: 0
```

Figure 19-3. The format menu from WordPerfect.

```
05-26-93  02:24p              Directory C:\WP51\*.*
Document size:   40,680    Free:  1,458,176 Used:   6,268,682    Files:       108

.     Current    <Dir>                  °  ..     Parent     <Dir>
LEARN    .       <Dir>    03-19-92 10:57a °  8514A    .VRS      5,226  03-09-92 12:00p
ALTB     .WPM         73  05-11-93 10:22a °  ALTC     .WPM         75  05-16-93 09:21a
ALTD     .WPM         67  03-19-92 07:03p °  ALTH     .WPM        128  06-30-92 07:26a
ALTI     .WPM         99  03-11-92 07:54p °  ALTM     .WPM        285  03-18-93 09:57a
ALTP     .WPM        105  08-28-92 01:25p °  ALTQ     .WPM        107  01-25-93 08:44a
ALTS     .WPM        257  06-24-92 08:59a °  ATI      .VRS     37,635  03-09-92 12:00p
CALC     .WPM      7,972  03-09-92 12:00p °  CHARACTR .DOC     47,008  03-09-92 12:00p
CHARMAP  .TST     39,271  03-09-92 12:00p °  CODES    .WPM     28,149  03-09-92 12:00p
CONVERT  .EXE    109,591  03-09-92 12:00p °  CURSOR   .COM      1,452  03-09-92 12:00p
EDIT     .WPM     10,826  03-09-92 12:00p °  EGA512   .FRS      3,584  03-09-92 12:00p
EGAITAL  .FRS      3,584  03-09-92 12:00p °  EGASMC   .FRS      3,584  03-09-92 12:00p
EGAUND   .FRS      3,584  03-09-92 12:00p °  ENDFOOT  .WPM      4,169  03-09-92 12:00p
ENHANCED .WPK      3,837  03-09-92 12:00p °  EPFX286E .PRS     11,746  03-19-92 11:05a
EQUATION .WPK      2,974  03-09-92 12:00p °  FASTKEYS .WPK      2,999  03-09-92 12:00p
FIXBIOS  .COM         50  03-09-92 12:00p °  FOOTEND  .WPM      3,829  03-09-92 12:00p
GENIUS   .VRS     12,885  03-09-92 12:00p °  GRAB     .COM     16,450  03-09-92 12:00p
GRAPHCNV .EXE    122,368  03-09-92 12:00p °  HPLASEII .PRS     59,587  01-14-93 10:02a
HPLASIII .PRS    145,300  07-04-91 12:09a    HRF12    .FRS     49,152  03-09-92 12:00p

1 Retrieve; 2 Delete; 3 Move/Rename; 4 Print; 5 Short/Long Display;
6 Look; 7 Other Directory; 8 Copy; 9 Find; N Name Search: 6
```

Figure 19-4. By pressing F5 (function key 5) in WordPerfect, the program gives a list of subdirectories and files in the current directory, with a menu at the bottom for selecting file handling commands.

Examples include setting margins, line spacing, page breaks, tab stops and many other features.

PRINTING

Printing means producing hard copy on paper from a document loaded and active in the word processor or a document stored on a disk.

HANDLING

Handling refers to the file management capabilities of the word processing application. By interacting with the computer's operating system, the word processor can copy, delete, move or rename files. Some word processors also have a view capability, allowing the user to look at a document without having to load the document into the word processor.

COMMON FEATURES OF WORD PROCESSORS

Although all word processors operate a little differently and with a few different functions, there are some features which are common to most.

CURSOR MOVEMENT

As mentioned earlier, the cursor indicates where the user is typing on the computer's screen. Word processors make use of the arrow cursor movement keys to allow the user to move the cursor up, down, right, and left. The page down and page up keys allow the user to move through a document one page (or screen) at a time. The home key usually moves the cursor to the beginning of a line. The end key moves the cursor to the end of a line.

Many programs also allow for more advanced movement. In WordPerfect, using the control key with the left and right arrow keys can move the cursor one word at a time. Using the home key with the up or down arrow keys will move the cursor one screen at a time. The control key plus the home key allows the user to specify which page in the document to go to. Pressing the home key twice plus an up arrow key moves the cursor to the first line of the document. Doing the same with the down arrow key moves the cursor to the last line of the document.

HELP

Most word processors have a help command or function key to provide the user with a summary of commands and their functions. Although it does not replace a user's manual, the help screen can sometimes remind the user of the capabilities that go with commands.

```
Up/Down Arrow

      Moves the cursor up/down one line.

If you press:                    The cursor will move to . . .
Home, Up/Down Arrow              The top/bottom of the screen.
Home, Home, Up/Down Arrow        The beginning/end of the document.
Home, Home, Home, Up Arrow       The beginning of the document before any codes.

Esc, n, Up Arrow ()              n lines up.
Esc, n, Down Arrow ()            n lines down.

GoTo, Up Arrow ()                The top of the current page or column.
GoTo, Down Arrow ()              The bottom of the page or column.




Selection: 0                                        (Press ENTER to exit Help)
```

Figure 19-5. A summary of the cursor movement capabilities of the up/down arrows in WordPerfect. (Taken from a WordPerfect Help screen.)

SCROLLING

As the document is typed, it appears on the screen. When the screen gets full, the computer stores information in memory and gives the user a clean, fresh area on which to type. *Scrolling* allows the user to view the information ahead and back, as well as sideways. Most video screens display 24 lines, with 80 characters or less on one line, depending on the size of text used. Most documents are over 24 lines long and some type sizes allow more than 80 characters per line. In either case, scrolling allows the user to see all of the document on the screen, although usually not all at one time.

DELETING/INSERTING

Text can be deleted in a word processor in several ways. Pressing the backspace key deletes text to the left of the cursor one character at a time. Pressing the delete key deletes the character under the cursor and moves the next character to the right to the position under the cursor. Most word processors have a *blocking* or marking function, that allows a section of text to be blocked so the entire block can be deleted.

Most word processors allow for two methods of typing. The typeover method allows the user to type over existing text, deleting the old text while typing the new text. The insert method allows the user to type in new text while pushing any existing text to the right. This feature is usually controlled by the insert key. Text may also be inserted into a document from another document file stored on a disk.

COPYING AND MOVING

The blocking feature mentioned above can also be used to move or copy a section of text to another location in the existing document or another document. Text may be blocked and moved using the mouse or a function key command and arrow keys.

SEARCHING/REPLACING

Searching is the ability of the word processor to search through a document looking for a word or *code*. This feature is especially useful for documents which will include a glossary. The user can insert a code, such as italics, for each word in the text to be included in the glossary. Then a search for the italics code will take the user to each word to be included in the glossary.

Another time-saving part of this feature is the *search and replace* capability. The user may decide to use one word instead of another, or may want to replace a capitalized word with lower case. The computer will search for the word and replace it. The user can usually choose to confirm each replacement or allow the computer to replace the word throughout the entire document without confirmation. For example, the word "operator" could be replaced with the word "user" throughout the text, or just in particular instances.

TEXT FEATURES

Commands that change the text from normal to text with special characteristics are grouped together and called *text features*. These features allow the user to emphasize certain parts of the text. Text features include changing size, underlining, capitalizing, boldfacing, superscripts, subscripts, and italics.

Although the word processor has the ability to create all or almost all of the features mentioned above, the printer may not. This is a common way in which the printer may limit the capabilities of the word processor.

DOCUMENT LAYOUT

Document layout refers to the commands used to position text and/or graphics in certain ways. These commands include centering, determining and changing margins, justifying margins, tab settings, paragraph indentions, setting spacing between lines and paragraphs and many others. In some programs, these settings are displayed in a "status line/bar" or "ruler."

```
L.....L.....L.....L.....L.....L.....L.....L.....L.....L.....L.....L.....L
|     ^     |     ^     |     ^     |     ^     |     ^     |     ^     |     ^
0"          +1"         +2"         +3"         +4"         +5"         +6"
Ctrl-End (clear tabs); Enter Number (set tab); Del (clear tab);
Type; Left; Center; Right; Decimal; .= Dot Leader; Press Exit when done.
```

Figure 19-6. The tab ruler in WordPerfect. Each "L" represents a tab stop, with an editing commands menu below the ruler.

CHECKING SPELLING

Most word processors come with a built-in spell checker. When the command is given, the word processor checks each word in the document to see if the word is included in its dictionary. This feature is very useful for poor spellers and/or poor typists. What the spell checker will not do, however, is check to see if the correct version of the word or the correct word is used. For example, when the user means to use the word "from" but instead types "form," the spell checker does not indicate that the word is incorrect. Likewise, when "too" is supposed to be used, but "to" is used instead, the spell checker will not catch it.

Some users get frustrated by the number of words the spell checker does not have in its dictionary, but this problem can be solved. When a word that is spelled correctly is not recognized by the spell checker, the operator generally has the opportunity to add it to the dictionary. (This feature should be used carefully.) Also, supplemental dictionaries with specialized words are available to increase the words in a particular user's dictionary. For example, an agricultural scientist may buy a special dictionary that includes correctly spelled words used in scientific classification of plants and animals.

PRINT OPTIONS

Most word processors offer the user a variety of print options. These include printing all or selected pages of a document, printing a document from a disk, selecting the quality of print, switching between types of printers, and printing or omitting graphics. These features allow the user to save time and computer paper.

ASCII FILES

Some word processors use *ASCII files* exclusively, but most save files

```
Print

    1 - Full Document
    2 - Page
    3 - Document on Disk
    4 - Control Printer
    5 - Multiple Pages
    6 - View Document
    7 - Initialize Printer

Options

    S - Select Printer                HP LaserJet III
    B - Binding Offset                0"
    N - Number of Copies              1
    U - Multiple Copies Generated by  WordPerfect
    G - Graphics Quality              Medium
    T - Text Quality                  High




Selection: 0
```

Figure 19-7. The print options menu available in WordPerfect.

created by the word processor as word-processor-text files, to maintain the various codes included in the formatting of the document. These word processors should also have the capability of importing or exporting files in what is known as ASCII or DOS format, with the specific codes of the word processor not included.

This feature is important because ASCII files can be viewed using the DOS TYPE command and, in the case of batch files, executed by the operating system. In addition, ASCII files can be transferred from one type of operating system to another without problems. Users who wish to type a document and then transport it over a network should use this feature.

THESAURUS

The *thesaurus* allows the user to mark a word and then obtain a list of synonyms (and sometimes antonyms) for a word. Most word processors with a thesaurus will allow the user to replace the marked word with one from the thesaurus by typing a corresponding number or letter.

The thesaurus for a particular program will be limited by the number of words/uses stored in the thesaurus file, but many programs have extensive capabilities. In addition, some programs offer specialized thesaurus files for different types of users.

UNDELETE

When a user deletes text, using either the delete key, the backspace key, or the blocking command, some word processors do not automatically delete it from memory. The information is stored in a buffer in RAM, and the user may recall the text using the *undelete* feature. Deleted text typically remains in the buffer until replaced with newly deleted information. Some word processors will keep three to five deletions separate in the buffer, with newly-deleted text replacing the oldest deletion.

ADVANCED FEATURES OF WORD PROCESSORS

While almost every commercial word processing program has some version of the common features mentioned above, some of the more popular (and more expensive) word processors have features that are advanced beyond standard features. Some of these are described below.

ADVANCED LAYOUT FEATURES

Advanced layout features include the ability to imbed *headers* and *footers*, *widow/orphan protection*, re-numbering pages in the middle of a document, WYSIWYG viewing of pages without having to print them, incorporation of graphics, changing paper sizes and types, and options about printing page numbers, to name a few.

```
Format: Page

    1 - Center Page (top to bottom)    No

    2 - Force Odd/Even Page

    3 - Headers

    4 - Footers

    5 - Margins - Top                  1"
                  Bottom               1"

    6 - Page Numbering

    7 - Paper Size                     8.5" x 11"
            Type                       Standard
            Labels

    8 - Suppress (this page only)


Selection: 0
```

Figure 19-8. The page format menu from WordPerfect. Note the ability to select different paper sizes, headers, footers, and page number positions.

GRAPHICS

The ability to import graphics into a document separates the most powerful word processors from the rest. This feature typically uses a graphics file that is separate from the document file and merges the two files when printing. Some programs have extensive graphics files that come with the program or that can be bought separately. Some have the ability to import a graphic generated by another program. WordPerfect features a grab utility that allows the user to copy a graphic from the computer screen and save it as a WordPerfect graphics file.

COLUMNS

The columns feature allows the user to type in two or more columns on the same page. The columns may be newspaper format, where the text in column one carries over to the next column, or parallel format, where the text in one column continues over to the next page in the same column.

Number of Cows:	34				
DIRECT EXPENSES	UNIT	PRICE	QUANTITY	DOLLARS	COST PER COW
INTEREST	DOLLAR	$8,700.00	8%	$696.00	$20.47
BULL	HEAD	$1,500.00	1	$1,500.00	$44.12
PASTURE	ACRE	$10.00	200	$2,000.00	$58.82
MINERALS	BAG	$10.00	16	$160.00	$4.71
PROTEIN MEAL	TON	$140.00	6.5	$910.00	$26.76
VETERINARY COW	HEAD	$15.00	34	$510.00	$15.00
MARKETING	DOLLAR	$0.04	11623	$464.92	$13.67
CHECK-OFF	HEAD	$1.00	31	$31.00	$0.91
GRAIN	TON	$145.00	7.5	$1,087.50	$31.99
VETERINARY CALF	HEAD	$8.00	31	$248.00	$7.29
VETERINARY BULL	HEAD	$15.00	2	$30.00	$0.88
REPAIRS/EQUIP.	YEAR	$1,000.00	1	$1,000.00	$29.41
TOTAL				$8,637.42	$254.03
Courtesy, M. Woods, Johnson Milling Company, Clinton, Mississippi.					

Figure 19-9. A table with a file imported from a spreadsheet.

TABLES

The tables feature automatically sets up tables with a specified number of rows and columns. A good word processor will allow the user to import a spreadsheet or database file into the word processor in column format. Some of the more popular word processors can import selected spreadsheets without losing any of the spreadsheet features. For example, in WordPerfect, a Lotus 1-2-3 file can be imported maintaining the same characteristics found in the spreadsheet. WordPerfect even keeps the formulas and relationships between parts of the spreadsheet intact.

GRAMMAR CHECKER

A *grammar checker* picks up the editing/proofreading function where spell checkers leave off. In addition to checking spelling, the grammar checker will check the meaning of words to see if they are used correctly in a sentence, check for passive writing, and grammatical and punctuation errors. Very few word processors have this feature built into the program. A couple of very good supplemental programs are available, however. Two of the most popular are Grammatik and RightWriter. RightWriter includes a copy of Strunk and White's *Elements of Style.*

MERGING DOCUMENTS

Merging documents requires the word processor to work like a database. A file containing fields of information can be merged with a file containing field variables to create multiple documents or pages without having to retype each one with just a couple of changes. The most common use is the merging of address and letter files. One file has each person's name and address while the other file may have a letter with field variables which are filled when the file is merged with the address file. (See Figure 19-10.) Some programs can be used in conjunction with a data base manager to accomplish the same result.

SPLIT SCREEN/WINDOWS

Use of a split screen or windows allows the user to work with more than one document at a time. In some programs, two documents can be loaded but only one shown on the screen at one time. A key is designated to switch back and forth between documents. Other programs allow the user to split the screen to see two (or more) documents at once. Word processors running under Microsoft Windows can open several documents at once.

MACROS

A *macro* is a file created with a set of commands or keystrokes that is combined into one keystroke to save the user time. Macros work on the same principle as batch files. In word processors, macros are generally used for commonly repeated keystrokes. For example, a user may create a macro

```
Database File (Secondary File)

Alicia Harvester{END FIELD}
133 Combine Road{END FIELD}
Farming Village, MS 39999{END FIELD}
{END RECORD}
------------------------------------------------
Johnny Appleseed{END FIELD}
Route 3{END FIELD}
Tractortown, OH 43210{END FIELD}
{END RECORD}
------------------------------------------------


Text File (Primary File)

                               November 21, 1993

{FIELD}1~
{FIELD}2~
{FIELD}3~


Dear {FIELD}1~:

     You have been selected to participate in the National Agriscience
Project Contest sponsored by American Farm Bureau................


Merged File (Page 1)

                               November 21, 1993

Alicia Harvester
133 Combine Road
Farming Village, MS 39999

Dear Alicia Harvester:

     You have been selected to participate in the National Agriscience
Project Contest sponsored by American Farm Bureau................
```

Figure 19-10. Parts of two files ready for merging are shown. The fields in the database file (secondary file) above are incorporated into the field variables in the text file (primary file).

with his or her signature line for letters to avoid typing in the signature line every time. Instead of typing "sincerely, name, and title" the user would type one or two keys to get the same result. In WordPerfect, a signature line might be stored in the macro file ALTS.WPM, and the user would type S while holding the Alt key to run the macro (which would print the signature line).

In addition to user-defined macros, some programs come with built-in macros containing commonly used sequences. These may include an on-screen calculator, print routines, and help screens. Users may use these "as-is" or edit them using a macro editor.

AUTOMATIC SAVE

In the setup menu of some word processors, the user can designate a set time when the word processor saves the current file to a disk. For example, if the user has designated 10 in the automatic save selection, then the word processor will save the current file after 10 minutes. This feature is very useful for those who forget to save regularly!

If a power loss occurs and the computer reboots, the word processor has the automatically saved file available for the user. This prevents losing work that wasn't saved. If the user exits from a document, the word processor erases the automatically saved file and restarts the clock when a new document is loaded.

```
Setup: Backup

     Timed backup files are deleted when you exit WP normally.  If you
     have a power or machine failure, you will find the backup file in the
     backup directory indicated in Setup: Location of Files.

        Backup Directory

     1 - Timed Document Backup                Yes
         Minutes Between Backups              15


     Original backup will save the original document with a .BK! extension
     whenever you replace it during a Save or Exit.

     2 - Original Document Backup             No




Selection: 0
```

Figure 19-11. The backup screen from WordPerfect's setup menu. Note that this user wishes to have the current file saved automatically every 15 minutes.

INDEX CREATION

Some word processors give the user the opportunity to create an index of important words used in a document. The program may do this in one of two ways. One way is to have the user mark the words in the document with an index code. At the end of the document, the user can give the index command and the word processor will generate an index that is alphabetized with page numbers after the words.

Another way is for the user to create a *concordance file* for the document. The concordance file contains a list of the important words. When using the index command, the user types the name of the concordance file and the program uses the words in the concordance file to generate the index. The first method is usually preferred because the concordance file method will list the page numbers of every occurrence of the words in the document, whereas the user may only want to index certain occurrences of the word.

THE WORD PROCESSED PAGE

A standard page of typing paper is 8.5 inches wide and 11 inches long. Using defaults of 10 *characters per inch (cpi)* and 6 lines per inch, the page will hold 66 lines of 85 characters each with no *margins*.

With standard default margins of 1-inch on the top, bottom, and each side, the page will hold 54 lines of 65 characters each. If 12 cpi characters are used, lines with a 1-inch margin on each side will hold 78 characters. (This is probably the justification for making a computer screen 80 characters wide.)

Other common margins are 1.5 inches for the left margin, especially when the printed pages are to be bound. Some style manuals require a bottom margin of 8 lines, or 1.33 inches.

PRINTER TYPEFACES, TYPESTYLES AND FONTS

Every computer printer comes with one or more built-in fonts. The word processor has drivers for different printers that allow it to work with the printer. When a different printer is selected, the word processor may have to change the document layout to fit the fonts available to the new printer. The following should help the reader understand what is meant by the term font.

- The typeface is a specific design for a set of letters. Common typefaces include Courier, Prestige, Times-Roman, and Helvetica. There are hundreds of different typefaces.

- A typestyle is a particular variation of a typeface. For example, a printer may have Prestige normal, Prestige bold, and Prestige italic.

- A *font* is a complete alphabet of a typestyle in one size. The size is usually designated in characters per inch (cpi) or point sizes. A point is equal to 1/72 of an inch, so a 12 cpi font is roughly equivalent to a 10 point font.

FONT DESCRIPTIONS

Various terms are used to describe fonts. One of these is serif, which refers to the extensions (or finishing strokes) on each letter in some typefaces. For example, Times-Roman has serifs. Sans-serif refers to a font without serifs. Helvetica is a sans-serif font.

```
Printer Fonts

Courier 10 cpi (normal)
```
`Courier 10 cpi (bold)`
`Courier 10 cpi (italic)`
`Prestige Elite 10 point (normal)`
`Prestige Elite 10 point (bold)`
`Prestige Elite 10 point (italic)`
Times Roman 12 point (normal)
Times Roman 12 point (italic)
Times Roman 14 point (bold)
Helvetica 12 point (normal)
Helvetica 12 point (italic)
Helvetica 14 point (bold)

Universal Scalable 18 point (bold, italic)

Universal Scalable 22 point (normal)

Figure 19-12. An example of selected fonts. Note that Helvetica and Universal Scalable are sans-serif fonts, while Courier, Prestige Elite, and Times-Roman have serifs.

Fonts may also be described based on letter spacing. A *fixed-pitch font* uses the same space on a page for each character, regardless of the actual space the character occupies. A *proportionally-spaced font* varies the space given to each letter based on the width of the letter.

Another way to describe fonts is by how they are developed and interpreted by the printer. In a *bitmapped font*, each character is described as a pattern of dots in a specific point size. The font must be available in a certain point size before it can be used. A *scalable font* has each character described as a mathematical formula. The user can designate any size type by changing the point size. Scalable fonts require the printer to have a *font renderer*, a scheme to interpret the mathematical formula into a printable character. For this reason, scalable fonts usually take longer to print than bitmapped fonts.

PAGE DESCRIPTION LANGUAGES

Printers are often described by the *page description languages* they can use to interact with software. Page description languages are programming languages that describe the placement and appearance of text and graphics on a page. For laser printers, the most common are Adobe PostScript, Hewlett-Packard Printer Control Language (HP PCL), and Hewlett-Packard Graphics Language (HP-GL).

Adobe PostScript was developed as a common language for describing pages to almost any output device, including printers, typesetters, and slidemakers. It has been the language of choice in the publishing industry for several years. Adobe PostScript delivers high-quality text and graphics and supports thousands of fonts and many software applications.

HP PCL was developed for HP LaserJet printers. The HP LaserJet III and IV use PCL 5, which is similar to PostScript. HP GL is a new language designed to improve the graphics capabilities of Hewlett-Packard printers. The combination of HP PCL 5 and HP GL give these printers a very flexible language.

CHAPTER QUESTIONS

1. How do word processing programs work?

2. What is the most popular word processing program in use today?

3. What type of computer would you buy to use primarily for word processing? Give specifications and justify your choices.

4. Briefly describe the five major functions of word processors.

5. Describe six common features of word processors.

6. Describe six advanced features of word processors.

7. Describe a typical page of a word processed document.

8. What is the difference between a typeface, a typestyle, and a font?

9. What is the difference between a bitmapped font and a scalable font?

Chapter 4

ELECTRONIC SPREADSHEETS/ RECORDKEEPING

The paper spreadsheet has been used in accounting for many years. It consists of a series of columns and rows, with space for column and row headings to keep track of accounts.

In 1979, Dan Bricklin and Bob Frankston introduced VisiCalc, the first electronic spreadsheet, designed to run on an Apple II computer. The idea was simple, to create an electronic version of the paper spreadsheet that had been a standard for decades. The introduction of VisiCalc is probably the most important reason that computers have become popular for businesses.

The flexibility and powerful features of the modern electronic spreadsheet continue to make it a popular choice for business. This chapter provides an overview of how electronic spreadsheets work and describes some of the most useful features of spreadsheets.

OBJECTIVES

1. Define electronic spreadsheets.

2. Describe the basic components of electronic spreadsheets.

3. Describe the major functions of spreadsheets.

4. Identify essential features of spreadsheets.

5. Associate standard spreadsheet commands with their functions.

6. Associate selected advanced spreadsheet commands with their functions.

7. Identify features of popular accounting software.

ELECTRONIC SPREADSHEETS: A FLEXIBLE RECORDKEEPING TOOL

An electronic spreadsheet—spreadsheet for short—is an all-purpose computer application that can be used for almost any task involving the organization of numbers. It is based on a paper spreadsheet with columns and rows. The user keys in the titles and numbers and any calculations that might be needed.

One advantage of the spreadsheet is that once it is set up properly, the user can save time by never having to set up the spreadsheet again. For example, suppose the user has an employee's schedule of time worked set up in a spreadsheet. When one month is over, the user can simply copy the spreadsheet to another file or another location on the same spreadsheet and enter the numbers for the next month. The spreadsheet will automatically recalculate where it is supposed to so the user will know how much to pay the employee, the amount of taxes to pay, etc. In a company with many employees, a blank schedule of time worked can be developed with all the titles and formulas and no data. This blank schedule is an example of a *template*. The template can be copied to each employee's spreadsheet file and then the data may be entered.

Another advantage of the spreadsheet is its capability of exploring "what-if" scenarios. Once the formulas have been entered into the spreadsheet, the user can change input information to see the results of possible changes without having to implement the changes. This feature is useful in budgeting, determining prices of products, and submitting bids for services.

TERMS

accounting software	formula	template
cell	label	worksheet
cell address	prompt area	value
companion programs	range	

ELECTRONIC SPREADSHEETS/RECORDKEEPING 105

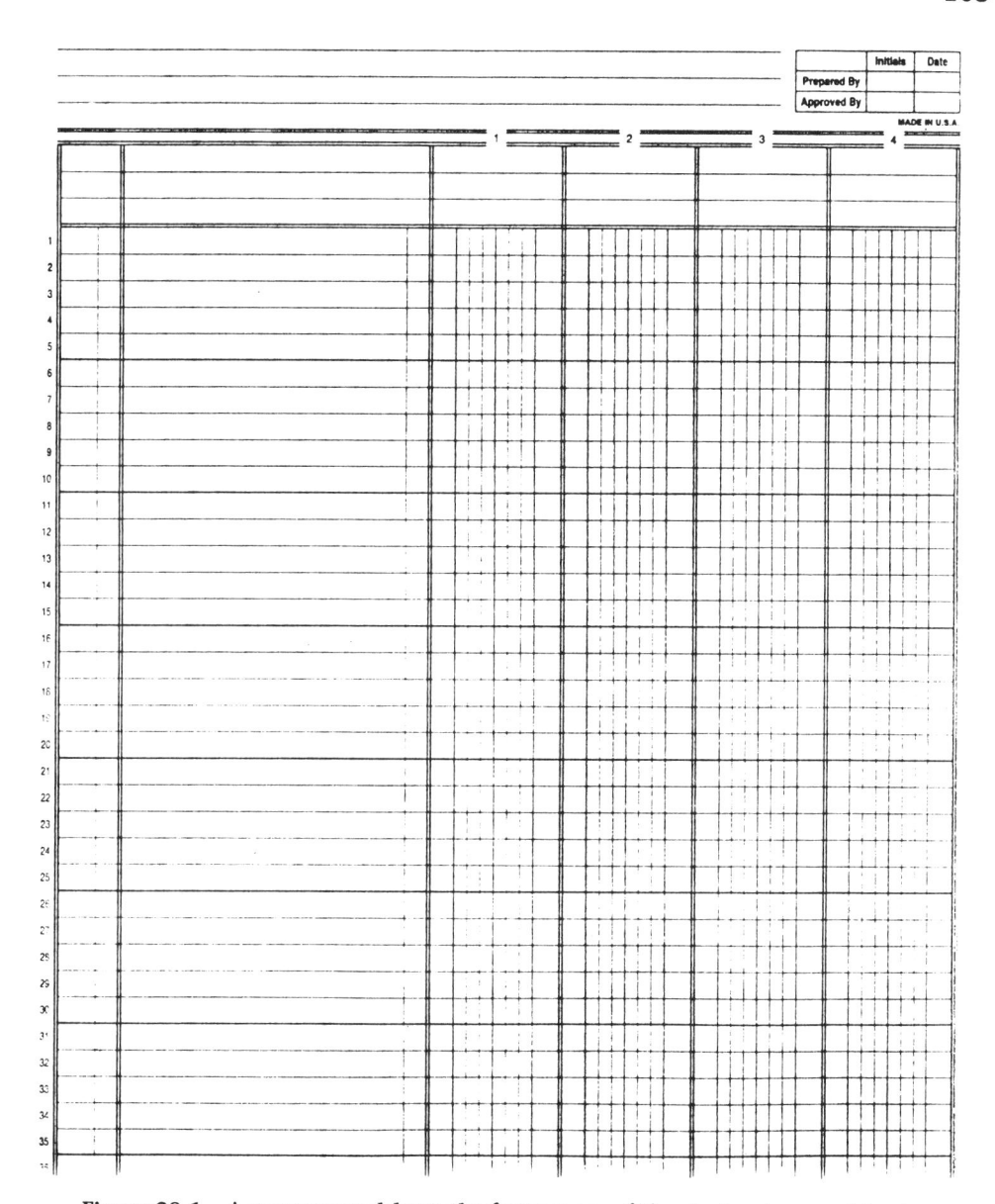

Figure 20-1. A paper spreadsheet, the forerunner of the electronic spreadsheet.

SPREADSHEET SOFTWARE

The most popular spreadsheet available today is Lotus 1-2-3, marketed by Lotus Development Corporation. Other popular spreadsheets with powerful features include Quattro Pro from Borland, Microsoft's Excel, and SuperCalc from Computer Associates.

Figure 20-2. Lotus 1-2-3 is the most popular spreadsheet on the market today.

Because the basic spreadsheet is a fairly simple application, packages such as InstaCalc from FormalSoft, Lucid 3-D from DacEasy and the shareware program PC-Calc provide users with a functional spreadsheet for a low price. These packages don't have all the features of the big packages, but provide a lot of program for the money. Table 20-1 contains some of the more popular spreadsheet application software packages.

HARDWARE REQUIREMENTS

Full-featured spreadsheet applications such as Lotus 1-2-3 and Quattro Pro will recommend 4MB of RAM and up to 10MB of hard disk space to run the latest versions. These programs will not function as intended on

Table 20-1
POPULAR SPREADSHEET APPLICATION SOFTWARE PACKAGES

Package	MS-DOS	Windows	Macintosh
Lotus 1-2-3	Yes	Yes	Yes
Quattro Pro	Yes	Yes	No
Microsoft Excel	Yes	Yes	Yes
SuperCalc	Yes	No	No
Lucid 3-D	Yes	No	No
InstaCalc	Yes	No	No
PlanPerfect	Yes	No	No

machines with no hard disk drive. Some of the less expensive programs will work on a computer without a hard disk and only 640K of memory. The smaller memory may, however, limit the amount of data that can be entered at one time.

Spreadsheet calculation times can be improved with the addition of a math coprocessor for machines using a microprocessor without a built-in math coprocessor (486sx or lower for DOS machines). The math coprocessor is not recommended, however, for users who have small worksheets or who plan to use the programs less than two hours per day. These users are unlikely to notice the time that is gained with a math coprocessor.

Spreadsheet programs designed to run under Microsoft Windows or on the Macintosh will benefit more from a laser printer than DOS-based programs. The Windows programs provide excellent font/graphic support in conjunction with the Windows software. Newer DOS versions of Excel, Quattro Pro, and Lotus 1-2-3 also provide fairly good printer and graphics support. For DOS-based programs, however, a good 24-pin dot matrix printer is probably the best choice for the money.

BASIC COMPONENTS OF ELECTRONIC SPREADSHEETS

When a user loads a spreadsheet application, the screen will be divided into two parts. The *prompt area* will take the top 3 or 4 lines of the screen. It contains the command line or ruler (accessed by pressing the slash [/] key), that shows the user which commands are available. It also contains

108 COMPUTER APPLICATIONS IN AGRICULTURE AND AGRIBUSINESS

Figure 20-3. Microsoft Excel, a popular spreadsheet package.

the entry/status line (two lines in some programs), that tells the user what is in a cell or shows what is being typed into a cell.

The *worksheet* area is divided into columns and rows. Columns are vertical areas extending from the top of the spreadsheet to the bottom. Columns are labeled with letters (column A, column B, etc.). Rows are horizontal areas extending from the left side of the spread sheet to the right side. Rows are labeled with numbers (row 1, row 2, etc.) A border around the worksheet area keeps the row and column labels visible at all times.

A *cell* is the place where a column and a row intersect. Cells are labeled by their column and row. For example, cell B22 is at the intersection of column B and row 22. The label B22 is an example of a *cell address*. This cell address is important, as it can be used to identify the contents of the cell.

ELECTRONIC SPREADSHEETS/RECORDKEEPING

Cells may have one of three types of information. Descriptive text used primarily to provide headings for the various rows and columns is referred to as a *label*. A number that is entered into a cell is referred to as a *value*. A *formula* is a mathematical expression or calculation entered into a cell. The number displayed as a result of the formula is called a variable.

MAJOR FUNCTIONS OF ELECTRONIC SPREADSHEETS

As noted earlier, spreadsheets are flexible programs that can perform almost any function requiring the organization and manipulation of numerical data. The data is organized in table format in a document called a worksheet. Like word processing programs, spreadsheet software has five major functions, which are described below.

CREATING

Creating involves starting with a blank worksheet and entering all of the labels, values, and formulas necessary to make a functional worksheet. For common operations, spreadsheet templates can be used to avoid some of the problems associated with developing worksheets from scratch.

The spreadsheet software allows the user to change formats such as

```
   Number of Cows:      34
   -----------------------------------------------------------------
   DIRECT EXPENSES   UNIT        PRICE   QUANTITY     DOLLARS   COST PER COW
   #################################################################
   INTEREST          DOLLAR    $8,700.00        8%    $696.00        $20.47
   BULL              HEAD      $1,500.00        1   $1,500.00        $44.12
   PASTURE           ACRE         $10.00      200   $2,000.00        $58.82
   MINERALS          BAG          $10.00       16     $160.00         $4.71
   PROTEIN MEAL      TON         $140.00      6.5    $910.00        $26.76
   VETERINARY COW    HEAD         $15.00       34    $510.00        $15.00
   MARKETING         DOLLAR        $0.04    11623    $464.92        $13.67
   CHECK-OFF         HEAD          $1.00       31     $31.00         $0.91
   GRAIN             TON         $145.00      7.5   $1,087.50       $31.99
   VETERINARY CALF   HEAD          $8.00       31    $248.00         $7.29
   VETERINARY BULL   HEAD         $15.00        2     $30.00         $0.88
   REPAIRS/EQUIP.    YEAR      $1,000.00        1   $1,000.00       $29.41
   -----------------------------------------------------------------
   TOTAL                                            $8,637.42      $254.03
   #################################################################
   Courtesy, M. Woods, Johnson Milling Company, Clinton, Mississippi.
```

Figure 20-4. A printout of a Lotus 1-2-3 worksheet containing an expense budget for a cow/calf operation. (Courtesy, Matt Woods, Johnson Milling Company,

column widths and types of numerical displays. Users can also insert columns and rows as needed to make a worksheet more functional.

REVISING

Revising refers to making changes in an existing worksheet. These changes may be in the format of the worksheet, in the values and labels entered, or in the formulas.

```
  Number of Cows:        38
------------------------------------------------------------------------
  DIRECT EXPENSES    UNIT        PRICE    QUANTITY     DOLLARS   COST PER COW
########################################################################
  INTEREST           DOLLAR    $8,700.00      10%      $870.00       $22.89
  BULL               HEAD      $2,000.00       1     $2,000.00       $52.63
  PASTURE            ACRE         $10.00     200     $2,000.00       $52.63
  MINERALS           BAG          $10.00      16       $160.00        $4.21
  PROTEIN MEAL       TON         $140.00     6.5       $910.00       $23.95
  VETERINARY COW     HEAD         $15.00      34       $510.00       $13.42
  MARKETING          DOLLAR        $0.04   11623       $464.92       $12.23
  CHECK-OFF          HEAD          $1.00      31        $31.00        $0.82
  GRAIN              TON         $145.00     7.5     $1,087.50       $28.62
  VETERINARY CALF    HEAD          $8.00      31       $248.00        $6.53
  VETERINARY BULL    HEAD         $15.00       2        $30.00        $0.79
  REPAIRS/EQUIP.     YEAR      $1,000.00       1     $1,000.00       $26.32
------------------------------------------------------------------------
  TOTAL                                                $9,311.42    $245.04
########################################################################
```

Courtesy, M. Woods, Johnson Milling Company, Clinton, Mississippi.

Figure 20-5. A revised version of the spreadsheet in Figure 20-3. Note the results of changes in the number of cows, the interest rate and price of the bull.

FORMATTING

As already mentioned, spreadsheet software makes it easy for the user to change the format of the spreadsheet. Commands are available to change column width, fonts, label placement, and numerical displays (decimal places, percentages, dollar signs, etc.).

PRINTING

Worksheets and graphics from worksheets may be printed to a printer, plotter, or disk. The printer is the most common output device. A plotter is sometimes used to produce color charts, but not commonly used to print the actual data in the worksheet. A spreadsheet may be printed to a disk as an ASCII file, where it can be used by a word processor or transferred to another computer over a network. The resulting ASCII file, however, does not include the formulas and other codes contained in the spreadsheet.

HANDLING

Spreadsheet software has file management functions for copying, deleting, moving, and renaming files. These functions are performed in conjunction with the operating system or environment (such as Windows).

ESSENTIAL FEATURES OF SPREADSHEETS

Although the basic operating principles of spreadsheets haven't changed much since VisiCalc, the popular spreadsheet packages continue to add features to make the programs more powerful and easier to use. Some features that are essential to all spreadsheets are discussed below. Some of the advanced features are discussed later in this chapter.

CURSOR MOVEMENT

In a spreadsheet, the cursor can be moved to different cells using the cursor control keys: the arrow keys, page up/page down keys, the home key, and the end key. Users often find it much easier to use a keyboard with dedicated cursor control keys—those not shared with a number pad. Because spreadsheets use a lot of numbers, most users wish to keep the number pad reserved for number entry only.

The other way to move around a spreadsheet is by using the "goto" command. Because each cell has its own address (its column and row), the user can give the goto command followed by the cell address to send the cursor directly to that cell. For example, in Lotus 1-2-3, the goto command is F5, followed by the cell address. To move to the cell at the intersection of column K, row 28, the user would press F5 and type K28.

SCROLLING

It is possible to create a worksheet that is much larger than the computer monitor's screen. For example, the Quattro Pro spreadsheet has a possible worksheet size of 230 columns (A through IV) and 8192 rows, a total of 1,844,160 cells! All of these cells couldn't possibly be shown on the screen at one time. Scrolling allows the user to see columns and rows not visible when the spreadsheet is in the home position.

So the user will know what the data in columns means, he or she can set titles of columns and/or rows that remain on the screen at all times. Some spreadsheets also allow the user to use multiple windows—to see two or more worksheets or two parts of the same worksheet at the same time.

DATA ENTRY

Text entered into a worksheet is referred to as titles or labels. Some spreadsheets require the user to put an apostrophe (') or quote (") symbol before the text to indicate that it is a label and not a value. (The apostrophe or quote does not show up on the worksheet, only on the entry line.)

Numbers entered into a worksheet are considered values. These values can be part of formulas entered in other cells of the worksheet. Formulas are mathematical expressions used to create variables. The user indicates to the spreadsheet that a formula is being entered by typing a left parenthesis symbol [(], an at sign [@], or a math symbol such as a plus sign [+]. After this symbol, basic mathematical rules apply, such as having to close parentheses after opening them. As with any mathematical formula, calculations within parentheses are performed before those outside, and division and multiplication are performed before addition and subtraction. Spreadsheets also have built-in commands, called @ commands. These make formula writing easier by reducing the number of keystrokes required to obtain the same answer. For example, if the user wants to add the values in cells E1 through E10, he or she could enter

+E1+E2+E3+E4+E5+E6+E7+E8+E9+E10

Table 20-2
COMMONLY-USED @ SPREADSHEET COMMANDS AND FUNCTIONS

Command	Function
@COUNT	Gives a count of the number of cells in a specified range
@DATE	Enters the current date in a cell
@EXP	Raises a number to a specified exponent
@MAX	Finds the maximum value in a range of cells
@MIN	Finds the minimum value in a range of cells
@SQRT	Finds the square root of a value or variable
@SUM	Adds the value of cells in a specified range
@TIME	Puts a clock in a cell

ELECTRONIC SPREADSHEETS/RECORDKEEPING

An easier way to do this would be to use an @ command called @SUM. To use this command, the user would enter

@SUM(E1..E10)

Both would give the same result. Table 20-2 contains a summary of commonly used @ commands.

Another data function is the built-in error warning given when a user enters a label (text) in a cell where the spreadsheet is expecting a value. The cell containing the formula will display "ERR" instead of the expected variable.

```
Number of Cows:         34
---------------------------------------------------------------------------
DIRECT EXPENSES    UNIT        PRICE    QUANTITY    DOLLARS      COST PER COW
###########################################################################
INTEREST           DOLLAR    $8,700.00       8%     +C5*D5       +E5/B1
BULL               HEAD      $1,500.00       1      +C6*D6       +E6/B1
PASTURE            ACRE         $10.00     200      +C7*D7       +E7/B1
MINERALS           BAG          $10.00      16      +C8*D8       +E8/B1
PROTEIN MEAL       TON         $140.00     6.5      +C9*D9       +E9/B1
VETERINARY COW     HEAD         $15.00      34      +C10*D10     +E10/B1
MARKETING          DOLLAR        $0.04   11623      +C11*D11     +E11/B1
CHECK-OFF          HEAD          $1.00      31      +C12*D12     +E12/B1
GRAIN              TON         $145.00     7.5      +C13*D13     +E13/B1
VETERINARY CALF    HEAD          $8.00      31      +C14*D14     +E14/B1
VETERINARY BULL    HEAD         $15.00       2      +C15*D15     +E15/B1
REPAIRS/EQUIP.     YEAR      $1,000.00       1      +C16*D16     +E16/B1
---------------------------------------------------------------------------
TOTAL                                              @SUM(E5..E16) @SUM(F5..F16)
###########################################################################
Courtesy, M. Woods, Johnson Milling Company, Clinton, Mississippi.
```

Figure 20-6. The same spreadsheet shown in Figure 20-4, with the formulas shown in cells rather than the variables resulting from the formulas.

STANDARD SPREADSHEET COMMANDS

Commands on almost every type of spreadsheet program can be accessed using the slash (/). Pressing the slash key will list the command menus in the prompt area. In many programs, the command ruler is shown on the screen at all times and can be accessed with either the slash key or a mouse. Most programs contain the following menus: worksheet, range, copy, move, file, and print. Others contain more menus which are discussed in a later section of this chapter.

WORKSHEET

The worksheet menu allows the user to make several changes to the entire worksheet. The following submenus are included in the worksheet menu.

Global

The global subcommand allows the user to make format changes to the entire worksheet. These include column width, column justification, and value characteristics such as number of decimal places, percentages, and dollar signs.

| Worksheet | Range | Copy | Move | File | Print | Graph | Data | View |

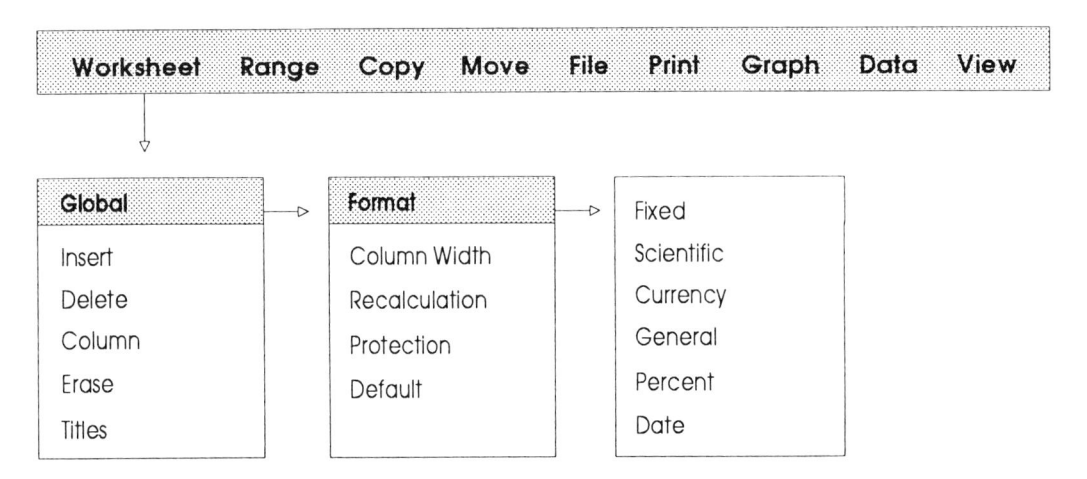

Global		Format		
Insert		Column Width		Fixed
Delete		Recalculation		Scientific
Column		Protection		Currency
Erase		Default		General
Titles				Percent
				Date

Figure 20-7. An abbreviated tree for the worksheet command and global subcommand in Quattro Pro for DOS.

Insert/Delete

The insert subcommand inserts a column or row at the cursor location or at another place specified by the user. The delete subcommand will erase a column or row.

Titles

The titles subcommand allows the user to specify a number of columns or rows that remain on the screen at all times. These serve as the titles for the scrolling columns and rows.

Erase

The erase subcommand is used to erase a worksheet from RAM. It does not erase the worksheet from the disk. Users should be careful to save any changes in the worksheet to a disk before using this command or the changes will be lost.

RANGE

The *range* command is similar to the global command. It allows the user to make format changes to a cell or group (range) of cells instead of the entire worksheet. The user can also name or erase a range of cells.

COPY

The copy command allows the user to copy a cell or range of cells to another location. One important feature is the ability of this command to adjust formulas to the different location. Some spreadsheets do this automatically, while others ask the user each time the command is used.

MOVE

The move command allows the user to move a cell or range of cells to another location.

FILE

The file command is the file management system of the spreadsheet. The user can retrieve a worksheet from a file, save a worksheet to a file, erase a file, copy a file, move a file, or combine two different files.

PRINT

The print command allows the user to print the spreadsheet. The user can specify the type of printer and range of cells to be printed. Certain columns and/or rows can be specified as borders for worksheets that span multiple pages when printed.

One option the user has is to print the spreadsheet to a disk in the form of an ASCII file. This is an important feature because the resulting ASCII file can be used in a variety of ways. When designating the filename to be printed to, the user should be careful not to use the same filename as the original worksheet, as it will save the ASCII file over the spreadsheet file. The user should use an extension, such as TXT or ASC, to indicate that the file is an ASCII file and not a spreadsheet file.

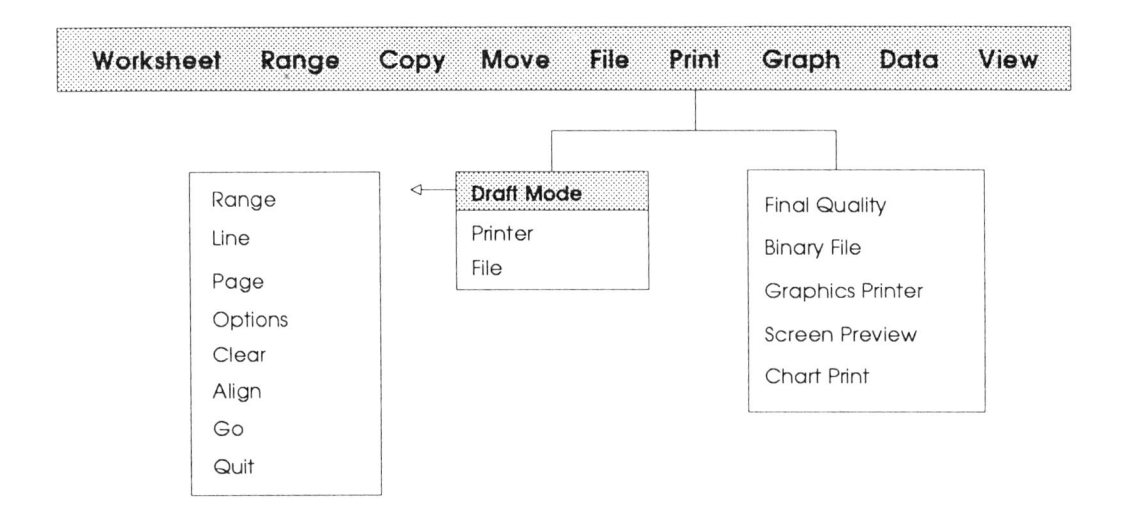

Figure 20-8. An abbreviated tree of the print command menu and printer submenus.

ADVANCED SPREADSHEET COMMANDS

While the standard features are similar on almost all spreadsheet programs, the more powerful programs include more options (submenus) under some of the standard commands and may also include more commands on the command line. Some of the more frequently used advanced commands are describe below.

WORKSHEET

In addition to the standard features of the worksheet command, some programs include advanced subcommands. The macros command allows the user to define and run macros to save time. (For a more detailed description of macros, see Chapter 19.)

RANGE

Advanced subcommands under the range command include the option to protect a range of cells and the ability to search for specific data in cells.

FILE

Although most spreadsheets have basic file management capabilities, some spreadsheets go a step further. An advanced subcommand under the file command gives the ability to import a file created by another application. Several spreadsheets can import files from popular database management programs such as DBase and Paradox. One spreadsheet, InstaCalc, can import files from Quicken, an accounting package discussed in the next section.

PRINTER

In addition to the print capabilities, some spreadsheets offer the view subcommand. View allows the user to see what the printout of a worksheet or graphic would look like without actually having to print it.

GRAPH

Early spreadsheets required *companion programs* to produce graphs from data entered into the spreadsheet. The ability to provide presentation-quality graphics is now a must for many spreadsheet users. Most spreadsheets have extensive graphics capabilities. Because of this, the graph command is often one of the commands on the main command line.

The graph command asks the user to select a graph type and indicate a range of cells to include in the graph. Another choice is for the user to enter plot values from the number pad or keyboard. The user can view the graph on screen without printing, and may be able to incorporate it into a presentation. Many options exist for customizing the colors and patterns of the graphs.

DATA

The data command assists the user in entering and sorting data. It includes a fill subcommand, which fills in between two cells with sequential data. For example, if the user indicates 100 for cell C1 and 1000 for cell C10, the program will fill in cells C2 through C9 with 200, 300, 400, etc.

Users can also use the data command to sort data and put it in table form for printing. Some programs include advanced statistics calculation as part of the data command.

VIEW

With the growth of GUIs, many DOS spreadsheets have been adjusted to make the interface of their programs more user-friendly. One of the newer features is the view command, which allows the user to change worksheet sizes, open multiple worksheets in individual windows, stack worksheets, and many other functions previously reserved for Macintosh or Windows users.

ACCOUNTING SOFTWARE

Users who do not wish to get involved in creating a spreadsheet system of recordkeeping may want to try accounting software. *Accounting software* is designed to be an easy-to-use alternative to electronic spreadsheets. Instead of setting up labels, titles, and formulas, the accounting package comes with a built-in set of accounts and the user enters only data into the existing framework.

The most popular accounting software package is Quicken from Intuit. Quicken comes with five basic accounts to let a person or business keep track of finances. The accounts are (1) cash, (2) checking, (3) credit cards, (4) assets and (5) liabilities. These accounts are functional but limited in scope.

Accounting packages do not have the flexibility of spreadsheet applications, but they work hard to provide customers with features that are commonly used in business in a format that is very user-friendly. For example, Quicken allows the user to write checks from the computer (using special checks available from Intuit) or make electronic payments using telephone hookups. Also, some of the more common financial analysis records, such as a cash flow statement or a net worth statement, are

ELECTRONIC SPREADSHEETS/RECORDKEEPING

Figure 20-9. Quicken, a popular accounting software package developed by Intuit.

computed and printed easily, whereas with a spreadsheet the user must set up these statements and write formulas to compute them.

Although accounting software is popular with some users, most agribusinesses with more than five employees will need to use a more flexible spreadsheet for keeping records. Larger agribusinesses may use software designed especially for their business by a computer programmer or obtain their software from their corporate offices.

CHAPTER QUESTIONS

1. What is the most popular spreadsheet program in use today?
2. What tool is the electronic spreadsheet based on? How are they alike? Different?

3. Define the following terms as they relate to electronic spreadsheets:

>   Cell
>   Address
>   Worksheet
>   Formula
>   Value
>   Label
>   Variable

4. Describe a good computer hardware system for operating a spreadsheet like Lotus 1-2-3 or Quattro Pro.

5. Briefly describe the five major functions of electronic spreadsheets.

6. Describe four standard spreadsheet commands.

7. Describe four advanced spreadsheet commands.

8. How are accounting software packages different from spreadsheets?

Chapter 5

DATABASE MANAGEMENT

Database management systems are probably the most flexible and powerful applications used on personal computers. The distinguishing characteristic of database management systems is the ability of the software to organize and manage both text and numerical data.

Agribusinesses use database management systems to perform a variety of tasks. Two important uses are inventory control and customer records.

Many people think database management systems are too difficult for novice computer users to operate. While it is true that these programs have many features, some complex and extensive, there are varieties of database software for all levels of users.

This chapter describes the basic concepts of database management. A look at some of the functions and features of the desktop versions of these powerful programs is provided.

OBJECTIVES

1. Define database management.
2. Identify functions of database management systems.
3. Describe database software and hardware requirements.
4. Discuss basic concepts associated with database management systems.
5. Describe features of database management programs.

DATABASE MANAGEMENT AS A BUSINESS APPLICATION

In general, a database is a collection of information called data. Database management is the use of the computer for maintenance, organization, and analysis of data.

Database management systems—database managers for short—are the software applications that allow users to manage data. Sizes of databases and programs vary greatly. Some commercial databases contain several gigabytes of information and have computer networks built around them. Others may be as small as a database of addresses and other information on 100 or fewer customers of a local feed store. This chapter focuses on the smaller databases that can be managed with desktop or personal computers.

FUNCTIONS

Database managers may be used for a variety of functions in an agribusiness. These flexible systems can be used to:

1. Generate mailing lists and labels
2. Generate directories of addresses
3. Maintain inventory control
4. Generate orders for supplies and other goods
5. Keep customer information on record
6. Maintain and analyze tax records
7. Generate work schedules
8. Organize and analyze production data

SOFTWARE

Database management software can be divided into two categories: single-purpose and general-purpose.

TERMS

Boolean "AND/OR" comparisons
database
field
Query-by-Example (QBE)
record
relational database
report
Structured Query Language (SQL)

DATABASE MANAGEMENT

Single-purpose database managers are designed to manage a specific database with specific types of data. These programs are generally easy to use but not very flexible. They can be large or small, and expensive or inexpensive. For example, shareware programs that organize mailing lists and fit on one 360K disk are available for a registration fee of $50 or less, while the database program used by the Dairy Herd Improvement Association is a very large (and probably very expensive) program.

The two types of general-purpose database software are further divided as programmable database managers and flatfile database managers. Both types are characterized by their flexibility that allows the user to customize the database and its management, but they vary by features and the amount of training necessary to operate the software. As a general rule, the more powerful the program, the more training required to familiarize the user with the many capabilities of the software. Most users agree that the amount of training is worth the time and expense because the operator is able to use the software properly.

Table 21-1
POPULAR DATABASE MANAGEMENT SYSTEM APPLICATION SOFTWARE

PACKAGE	MS-DOS	WINDOWS	MACINTOSH
dBASE	Yes	No	No
Paradox	Yes	Yes	No
FoxPro	Yes	Yes	No
PC File	Yes	No	No
Alpha Four	Yes	No	No
Q and A	Yes	No	No
PFS: File	Yes	No	No
R:BASE	Yes	No	No
Superbase	Yes	Yes	No
Omnis	Yes	Yes	Yes
Ingris	Yes	Yes	Yes

The most popular database manager program in use today is dBASE—a general-purpose, programmable database manager. Newer versions of programmable database managers such as dBASE, Paradox, and FoxPro have improved their interfaces in an effort to become more user-friendly, but have so many features that some training is commonly required before they can be used to their full potential.

Flatfile database managers are typically easier to operate but don't have as many features as programmable database managers. These programs still require the user to set up, or design the database, allowing for maximum flexibility in how the information is organized. Popular flatfile database managers include Alpha Four, Q and A, PFS:File, and PC File.

HARDWARE

The primary hardware concern associated with database managers is the amount of disk storage space and RAM they require. The minimum storage and RAM are determined by the size of the databases (the information) that need to be managed by the system. A hard disk drive is necessary to run the popular database managers. Most of the popular programs will run with 1MB of RAM, but more may be necessary, depending on the size of the database to be managed.

DATABASE MANAGEMENT CONCEPTS

Database managers organize information into two categories: fields and records. A *field* is a category of information such as name, account number, or zip code. A *record* is a collection of all fields for a subject, business or individual. Some database managers come with templates (pre-designed formats for recording information) for specific uses, such as customer or employee records.

A *database* is a collection of all the individual records which are kept in a file. This information may be printed to the screen, a printer, or to a disk file in the form of a *report*.

A *relational database* is a database with common data between fields and records. In other words, all records contain the same fields and all fields are located in all records. These fields may be linked to fields in another database. Linking means that updating the information in one database automatically updates the information in the related database.

The most powerful database managers use *Structured Query Language (SQL)*, a totally relational database language. Most of the popular programmable database managers used on personal computers use *Query-by Example (QBE)*, which has some relational characteristics. The low-end, flat file databases do not use a query language, but rely on menus and key words to search for specified fields and records.

SUMMARY OF DATABASE MANAGEMENT FEATURES

All database managers have the following in common:

1. The process involves keeping the same type of information on multiple units (also called subjects or individuals)

2. With a large number of units, the information can be managed more easily and efficiently using the computer than by hand

3. The database management system allows for updating, editing, and sorting the data contained in the database

Database managers differ in the number of features they offer the user and in the commands which access these features. Some of the more commonly used features of database managers are described below.

CREATING

Creating involves three processes. The first is designing the database—deciding what information the user wishes to maintain about each unit to be included in the database, how much space will be needed for each field, the type of field needed, and the organization of the fields in a record. The second step is defining the database—entering the design into the computer. The third step is entering the data into the database for each individual unit.

Field Names

For most database managers, field names must be 10 characters or less in length. A good rule of thumb is to keep the field name as short or shorter than the number of characters allocated to the data in the field. This keeps the field name from taking up more space in a listing than the data.

An example is the field "state," referring to the state in which a person lives, such as Ohio or Mississippi. State names can be long or short, but each state has a two-letter postal abbreviation. Since mailing labels require this abbreviation anyway, it is a good idea to keep it in the database in that form. Of course, a database field name of "STATE" would be longer than the postal abbreviation. Most users shorten the name of this field to "ST."

Types of Fields

In order to do a better job of managing data, some database managers require the user to define the field types when defining a database. A brief description of the field types follows. These descriptions show why defining field types can be useful.

Character fields. Character fields are used to store any characters such as letters and numbers.

Date fields. Date fields are used to store dates. The format is usually MM/DD/YY, requiring a field eight characters in length.

Numeric fields. Numeric fields store numbers. The numbers may contain a decimal point or a negative sign. If a field is to be used in a calculation, it must be defined as numeric.

Logical fields. Logical fields refer to fields that answer a question. Logical fields contain only one space and must be in the form of Y for yes, N for no, T for true, or F for False.

Memo fields. Memo fields are used to store large blocks of text and generally are not used to search or sort records. Some programs refer to these fields as global fields. These can be the largest fields in the database—some programs allow these files to be several thousand characters in length.

A Database Definition Example

The manager of a local farm supply business may determine that the following fields are needed for a customer database: last name, first name, farm name, address, city, state, zip, telephone, crops/livestock grown (beef, corn, cotton, etc.), sales, account number, account balance, credit rating, and comments about the customer. Table 21-2 contains a possible organization of these fields, their type, and their widths.

SORTING

One requirement of database management is the ability to sort the

Table 21-2
POSSIBLE FIELDS AND CHARACTERISTICS FOR A CUSTOMER DATABASE.

DESCRIPTION	NAME	TYPE	WIDTH
Last name	LNAME	Character	15
First name	FNAME	Character	15
Farm name	FARMNAME	Character	30
Mailing Address	MADDRESS	Character	30
Delivery Address	DADDRESS	Character	30
City	CITY	Character	16
State	ST	Character	2
Zip Code + 4	ZIP	Character	10
Telephone	PHONE	Character	12
Account Number	ACCTNUMBER	Character	10
Account Balance	ACCTBAL	Numeric	10
Last year's sales	SALES	Numeric	10
Current credit rating	CREDRATE	Numeric	10
Produce beef cattle	BEEF	Logical	1
Produce catfish	CATFISH	Logical	1
Produce corn	CORN	Logical	1
Produce cotton	COTTON	Logical	1
Produce dairy	DAIRY	Logical	1
Produce grain sorghum	GRAINSORG	Logical	1
Produce soybeans	SOYBEAN	Logical	1
Produce sheep	SHEEP	Logical	1
Produce swine	SWINE	Logical	1
Customer Comments	COMMENTS	Memo	800

Note that some fields contain numbers but are not listed as character fields. This is because those fields will not be used in any calculations.

information into almost any sequence. This is generally accomplished by specifying a field or fields by which to sort. The primary option is choosing ascending or descending order for the sort.

SEARCHING

Searching involves locating a record or records based on the information contained in a field or combination of fields. For example, the manager of the local farm supply business mentioned in Table 21-2 may want to send information about a new cotton growth regulator to customers who are cotton farmers. A search of the COTTON field in the customer database will let the manager know which of the customers grow cotton, then select these records and print their addresses.

Most database managers allow users to search for and select records by field contents using *Boolean "AND/OR" comparisons*. The Boolean "AND" means that the search will include every unit with a certain value in one field **and** a certain value in another field. The Boolean "OR" means that the search will include every unit with a certain value in one field **or** a certain value in another field. The "OR" comparison will yield more total units using the same fields and values.

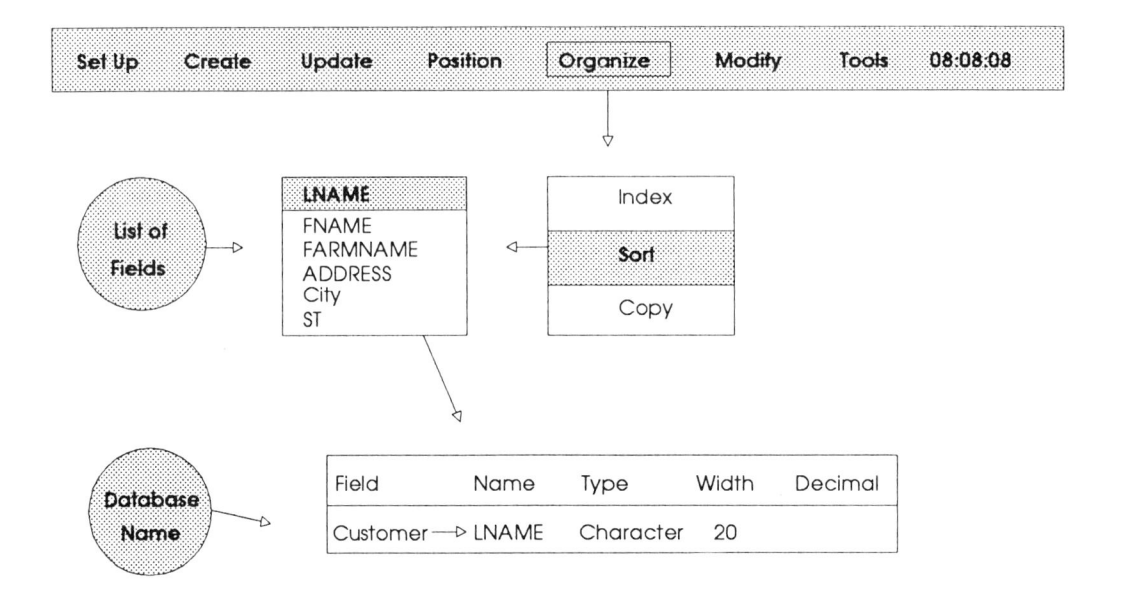

Figure 21-1. An abbreviated menu tree from dBASE III showing selected options under the organize command. The file is the CUSTOMER database defined in Table 21-2.

REPORTING

Reporting involves printing the information from all or selected fields in the database. Options include printing to the disk, printer or screen. The user may use pre-defined report formats that are included with the program or define his or her own report format. Defining a report format includes selecting the fields to be printed, selecting the records to be printed, and determining the organization of the selected records.

Boolean "AND/OR" Comparisons

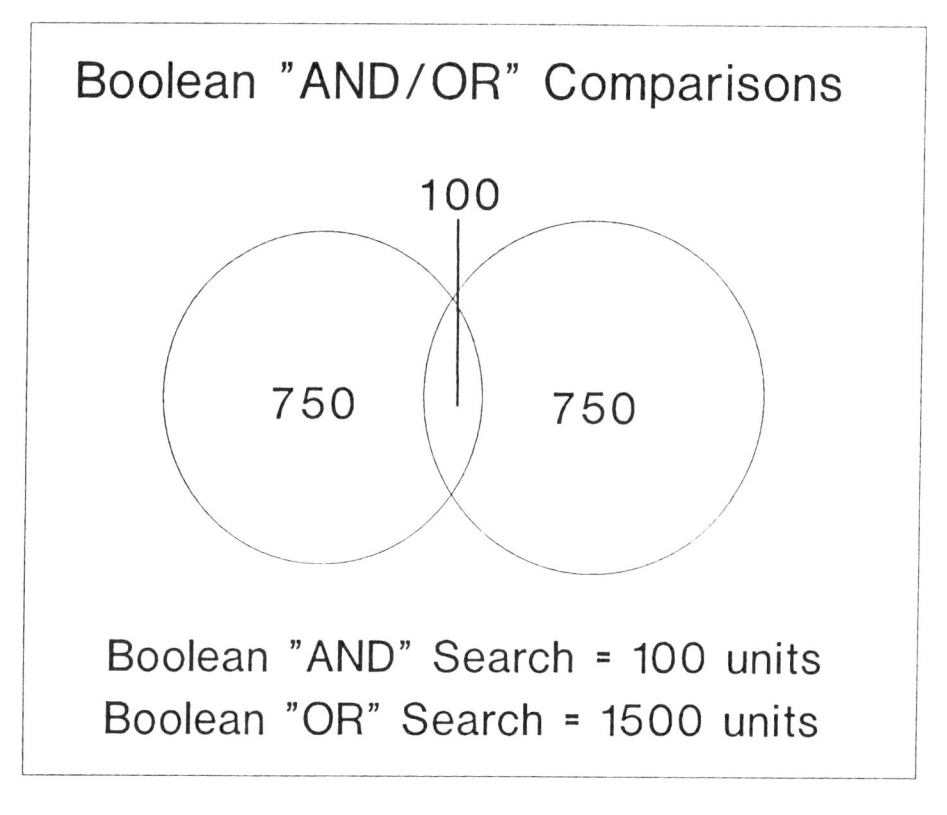

```
         100
    ___/___
   /   |   \
  /    |    \
 | 750 | 750 |
  \    |    /
   \___|___/
```

Boolean "AND" Search = 100 units
Boolean "OR" Search = 1500 units

Figure 21-2. A diagram showing the effects of Boolean "AND/OR" comparisons. 1,500 units meet one of the two search criteria values, but only 100 meet both.

UPDATING

Updating means changing the data in a database. This may involve changing data for existing records (either all or selected) in certain fields. It may also involve adding records to the database, deleting obsolete records, or undeleting deleted records.

CALCULATING

Most databases give the user the option of performing arithmetic calculations on fields with numerical data. These calculations may include summing data fields, computing an average of the fields, and counting the number of records with particular information. Most programs will also give subtotals of numerical fields based on classifications by a user-selected field.

```
B: Customer  150 records, disk can hold approximately 275 more
┌─────────────────────────────────────────┬──────────────────────────────────────────────┐
│ F1  ADD  - Add a record                 │ (ALT) F1  BIN  - Set binary search on/off    │
│ F2  MOD  - Modify a record              │ (ALT) F2  GLO  - Global update or delete     │
│ F3  DEL  - Delete a record              │ (ALT) F3  KEY  - Set up the smart keys       │
│ F4  DIS  - Display a record             │ (ALT) F4  NAM  - Alter field name or mask    │
│ F5  FIN  - Find a record                │                                              │
│ F6  LIS  - List or clone                │                                              │
│ F7  SOR  - Sort the Index               │                                              │
│ F8  UTI  - Utilities                    │                                              │
│ F9  MEN  - Show Smart Key Menu          │ (ALT) F9  END  - Quit or change database     │
└─────────────────────────────────────────┴──────────────────────────────────────────────┘
```
Please press the appropriate function key, or supply one of the 3-character commands ▶ _ ◀

Figure 21-3. Easy to use! The menu screen from PC File, originally a shareware program from Buttonware that has become commercial software. It is a very popular, inexpensive database manager.

MERGING/COPYING

The ability to merge databases or copy records from one database to another can save the user time and trouble. Time is saved because the data for each individual does not have to be entered again. Trouble is saved by avoiding keystroking errors that inevitably occur when entering data. This feature is sometimes called "cloning."

INDEXING

Indexing involves marking selected words in records and saving those words in an index file. The index file contains a list of the words and the number of each record that contains the words. The user can search using the indexed words to determine which records to select for an operation.

PROGRAMMING

With some database managers, programming refers to the creation of programs that act as macros do in word processors or spreadsheet applications. The programs are used to save the user time by combining a series of frequently used command sequences into one program that can be run using just one or two commands. Database programs vary greatly in this capability. Some programs offer basically no programming or macro capabilities; others, such as dBASE and Paradox, have their own programming languages, used for application development.

CHAPTER QUESTIONS

1. Select an agribusiness in your community and describe how that agribusiness might use a database management system.

2. What is the difference between a single-purpose database manager and a general-purpose database manager?

3. Define the following terms as they relate to database management:

    database
    field
    record
    relational database
    report

4. What is the difference between a Boolean "AND" search and a Boolean "OR" search?

5. Select five features of a database management system and describe what they do.

Chapter 6

GRAPHICS

The ability of users to generate graphics images using computers has increased dramatically in the last few years. Faster computers, better-quality monitors, and improvements in software have made it easy to produce quality graphics from personal computers. Items that had required the services of graphic artists and professional typographers are now being produced in-house by many agribusinesses.

The result of these new capabilities has been to allow agribusinesses to gain more outputs with the same personnel. The new packages are easy-to-use and require a minimal amount of training time and money. Most DOS graphics programs are menu-driven; programs designed for use under Microsoft Windows or on the Macintosh take advantage of the graphical orientation of their environments.

This chapter looks at the five major types of software used to produce graphics output. The features of each are described, with the primary focus given to presentation graphics programs, the ones that can do almost anything graphical.

OBJECTIVES

1. Describe the use of computer graphics in business.
2. Describe the different types of graphics software.
3. Describe the capabilities of presentation graphics software.
4. Discuss hardware requirements for various graphics functions.

COMPUTER GRAPHICS IN BUSINESS

Information that is presented in the form of a chart, graph, *line art*, or photograph is referred to as graphics. The term *business graphics* refers to using these graphics to convey this information to clients, management, shareholders, and others in a meaningful way. The use of computer software to generate business graphics with special effects designed to impress the viewer is referred to as *presentation graphics.* Presentation graphics include the special effects of color, patterns, and illustrations designed to "spruce up" a presentation.

The five most common types of graphs/charts produced in the business world are *pie graphs, line graphs, bar graphs,* line art/drawings, and pictures. Pie graphs, line graphs, and bar graphs are usually constructed by the software package after the user enters the data for the different parts of the graph, but the data may also be imported from a spreadsheet or database file. Line art can be drawn by the user or loaded into the software package from a file on a disk. Pictures may be loaded from files on a disk, from a CD-ROM disk, or from a paper copy using a scanner and scanning software.

Several standards have been developed for graphics files, especially for line art and illustration files. Users can buy floppy disks or CD-ROM disks filled with images produced in these *standard graphics file formats* for use with their software. Some of these standard graphics formats include: .PCX, .PIC, .DRW, .DXF, CGM and TIFF formats. Other formats have been developed to work with particular software and/or hardware. These include: .WPG (for use with WordPerfect), WMF (Windows Metafile), PICT (for use with MacDraw), HPGL (for use with the Hewlett Packard Graphics Language for HP printers).

TERMS

bar graph	floating-point calculations	portrait orientation
bullets	graphics drivers	presentation graphics
business graphics	landscape orientation	scale
computer aided design (CAD)	line art	screenshow
	line graph	standard graphics file format
film recorder	pie graph	

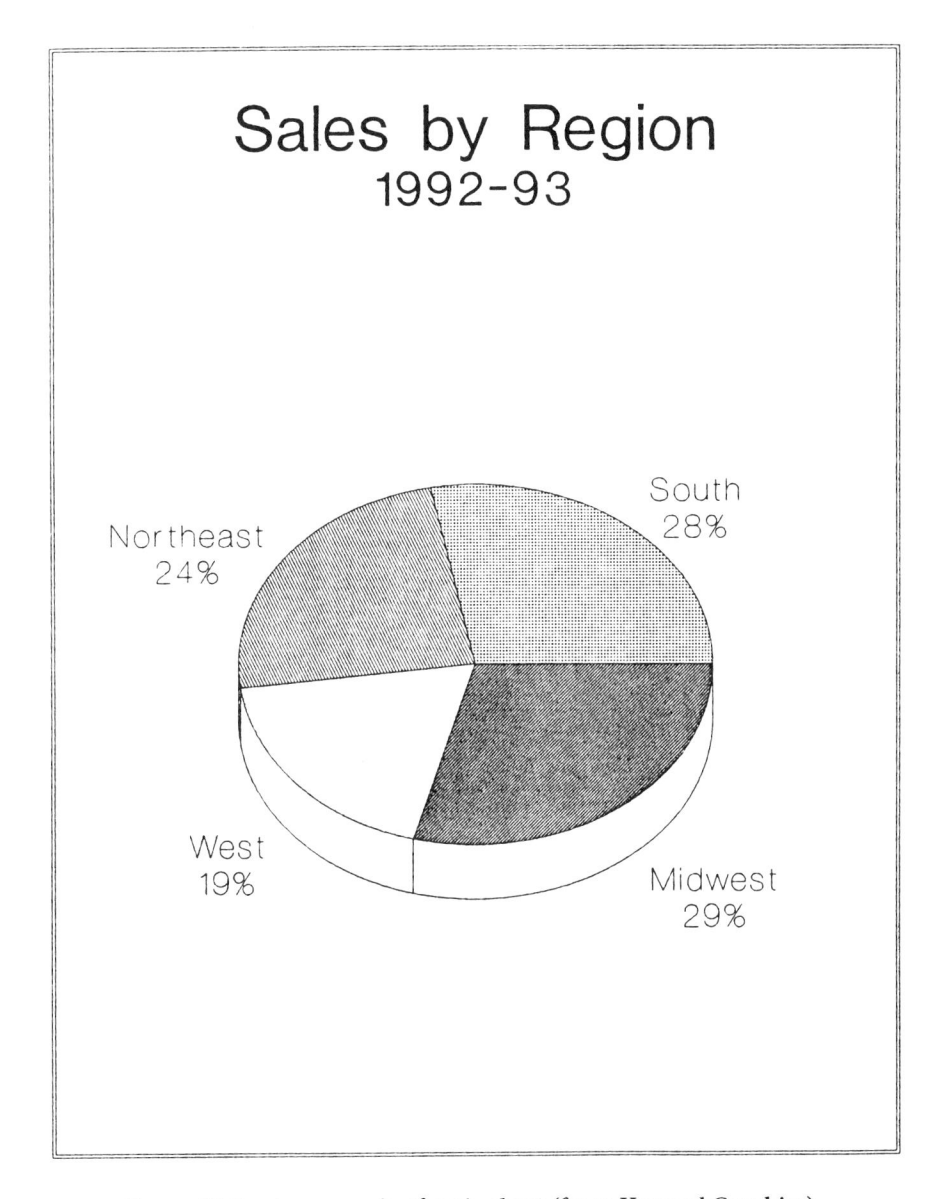

Figure 22-1. An example of a pie chart (from Harvard Graphics).

TYPES OF GRAPHICS SOFTWARE

Many different types of graphics programs are available. These programs are usually grouped by their primary function, although many have other capabilities as well. The five major types of graphics software are painting programs, drawing/drafting programs, canned graphics programs, presentation graphics programs, and other programs that incorporate graphics functions as an additional feature. Each of these types of programs are described briefly below. In the next section, the features of presentation graphics programs are discussed in more detail.

Figure 22-2. Good documentation is very important to the successful use of software. One argument for buying popular software is that easy-to-understand books are often available to supplement the user's manual.

PAINTING/DRAWING PROGRAMS

Painting/drawing programs are primarily designed for people with some artistic ability, although almost anyone can produce a decent looking product using these programs. These programs are characterized by their use of a mouse and icons representing different drawing tools, such as geometric shapes, pens, erasers, and paintbrushes. The user selects a pattern and color from the menu to fill in the designs. Text can be incorporated into the drawing as well. Once the user enters the text, it can be placed anywhere on the drawing screen.

GRAPHICS

Painting/drawing programs often allow the user to import line art or illustrations in one or more of the standard formats. The user can then make changes as desired. The output of these programs is often incorporated into other documents using one of the other types of graphics packages. Some programs can export files to make color 35mm slides on a *film recorder*.

Examples of popular painting/drawing programs include MacPaint, CorelDRAW, PaintShow, and Paintbrush. Programs that operate in Microsoft Windows or on a Macintosh are generally considered easier to use and usually interact with peripherals better because of the advanced *graphics drivers* in these programs.

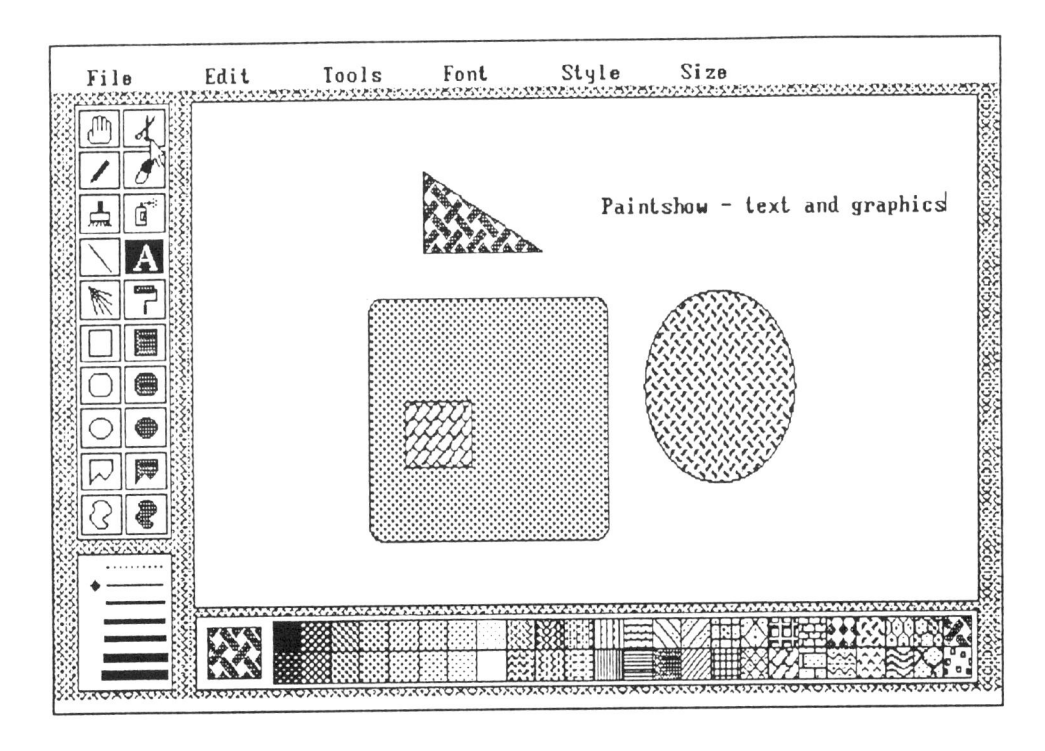

Figure 22-3. The Paintshow program with some examples of shapes and patterns. Note the icons for drawing tools on the left and for colors and patterns on the bottom. The pull down menu bar at the top makes the program easy to use.

CAD PROGRAMS

CAD programs are used to create original technical drawings. CAD stands for computer aided design. These programs are commonly used by

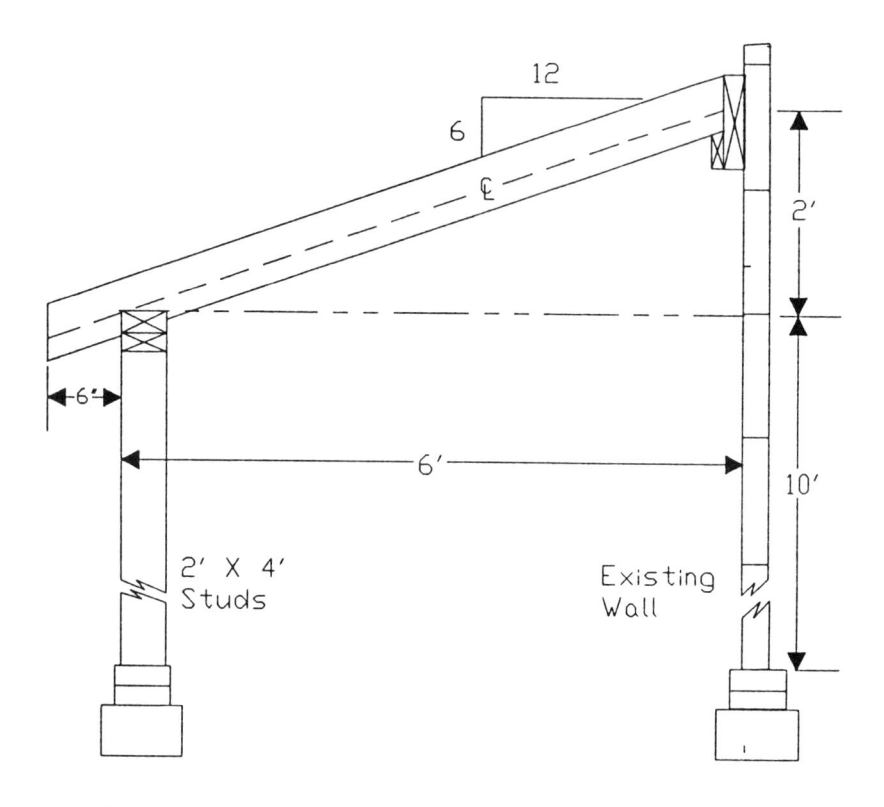

Figure 22-4. A drawing completed using a CAD program. (Courtesy, Donald M. Johnson, Mississippi State University)

interior designers, architects, landscape architects, engineering designers, and others.

These programs use mathematical formulas to draw blueprints and schematics, and diagrams to *scale*. The designs are saved to a disk, making revisions much more easily accomplished than with hand-drawn designs. Design databases are created so designers can easily search for designs to fit specific situations.

An important feature of CAD programs is the ability to include 3-D views, reverse angles, measurements, and other features without having to draw the designs separately. Some firms report a time savings of over 50 percent per designer when using CAD programs.

AutoCAD is the most popular CAD program. Other popular programs

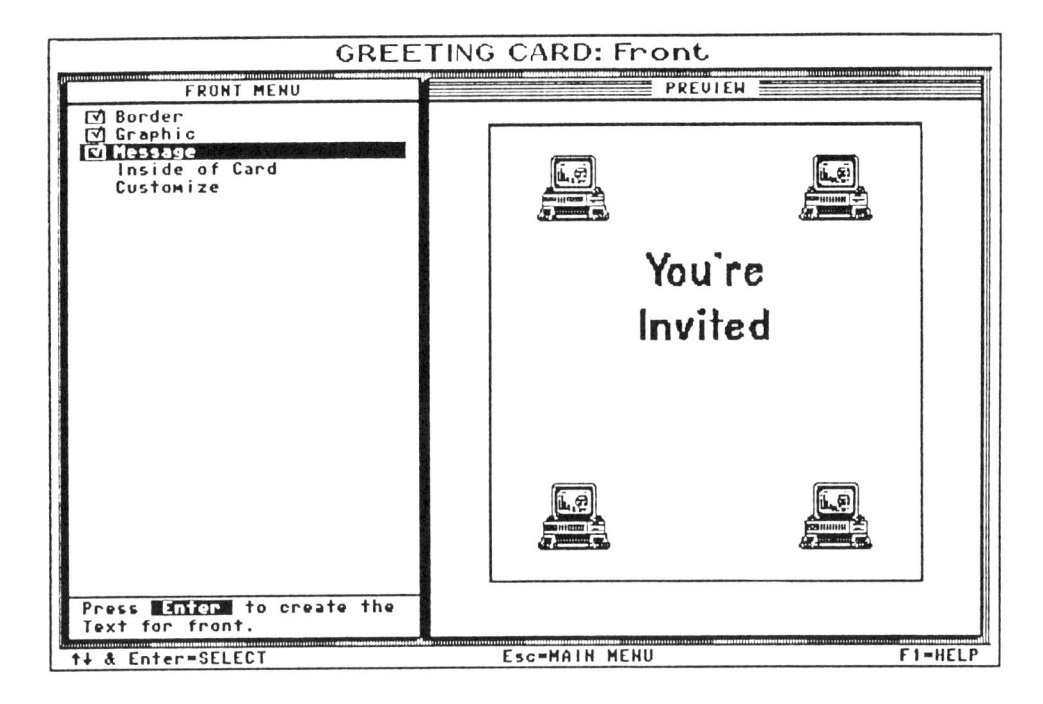

Figure 22-5. The greeting card production screen from Printshop, a canned graphics program.

include Generic CADD, MicroStation PC, VersaCAD, EasyCAD, AutoSketch, and FastCAD.

CANNED GRAPHICS PROGRAMS

Canned graphics programs feature an easy-to-use interface with a limited number of output capabilities. Line art can be imported into the program from disk files to improve the looks of the documents produced. Text and graphics can be incorporated into the same document.

Common outputs of canned graphics programs include banners, letterhead, posters, and cards (printed on one sheet and then folded by the user). Printshop, Print Master Plus, and PrintPARTNER are popular programs.

PRESENTATION GRAPHICS PACKAGES

Presentation graphics packages include several features in one graphics program. Most allow the user to draw figures, create various graphs using text and/or data, produce screen shows and 35mm slides, and much more.

Popular DOS versions of presentation graphics packages include Harvard Graphics, Freelance Graphics for DOS, DrawPerfect, and Applause II. Popular Windows versions include Microsoft PowerPoint, CA-Cricket Presents, Persuasion, and CorelCHART. For Macintosh computers, MacDraw and PowerPoint are popular packages.

OTHER GRAPHICAL PROGRAMS

Many other programs have the capability of including graphics in their regular outputs. The three primary types of software that can do this are desktop publishing packages, spreadsheets, and word processors.

Desktop publishing packages are designed to produce professional quality documents on a personal computer. These packages incorporate text and graphics into high-quality newsletters, brochures, and other documents that otherwise would be taken to a printing company. The quality of the printer and other output devices affects the quality of work that can be produced with these packages.

Spreadsheets that can produce graphics are rapidly becoming a standard in the business world. Quattro Pro was the first spreadsheet to incorporate the ability to produce presentation-quality graphics as part of the standard spreadsheet package. Windows-based spreadsheets have the ability to produce graphics that are comparable to those produced by the presentation graphics programs mentioned earlier.

Some of the high-end word processors, like Microsoft Word for Windows, Ami Professional, and WordPerfect, have included graphics support that allows them to function almost like desktop publishing packages. These programs include a standard library of line art that can be called up from a disk. Many painting/drawing packages and scanning software packages also have the capability of producing graphics that can be used by these programs.

FEATURES OF PRESENTATION GRAPHICS SOFTWARE

As mentioned earlier, presentation graphics software packages have numerous features. Some of the most important features are graph production, text charts, page orientation, drawing, importing graphics files, im-

GRAPH PRODUCTION

Graph production refers to the package's ability to produce the three common types of graphs: pie graphs, line graphs, and bar graphs. The most

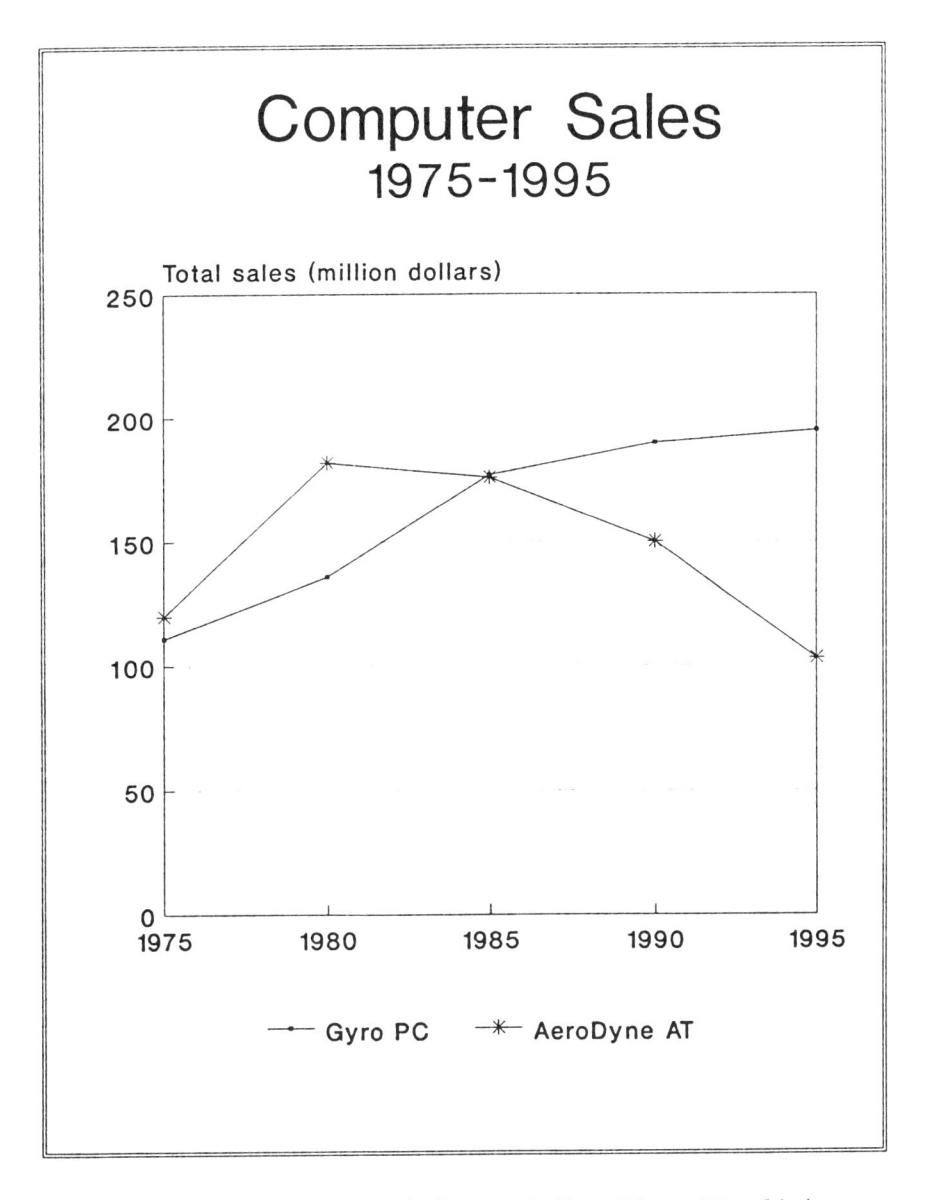

Figure 22-6. An example of a line graph (from Harvard Graphics).

standard way these are produced is for the user to specify titles and labels and then enter the data. Another way is to import data from a file on disk.

Pie charts are used to show parts of a whole and to make simple comparisons. Pie chart options include use of colors, patterns, comparison pies (two on the same chart), 3-D views, and breaking out of one or more pieces of the pie. Line charts are used to show trends and make simple comparisons. Bar charts are used to show trends, make simple comparisons,

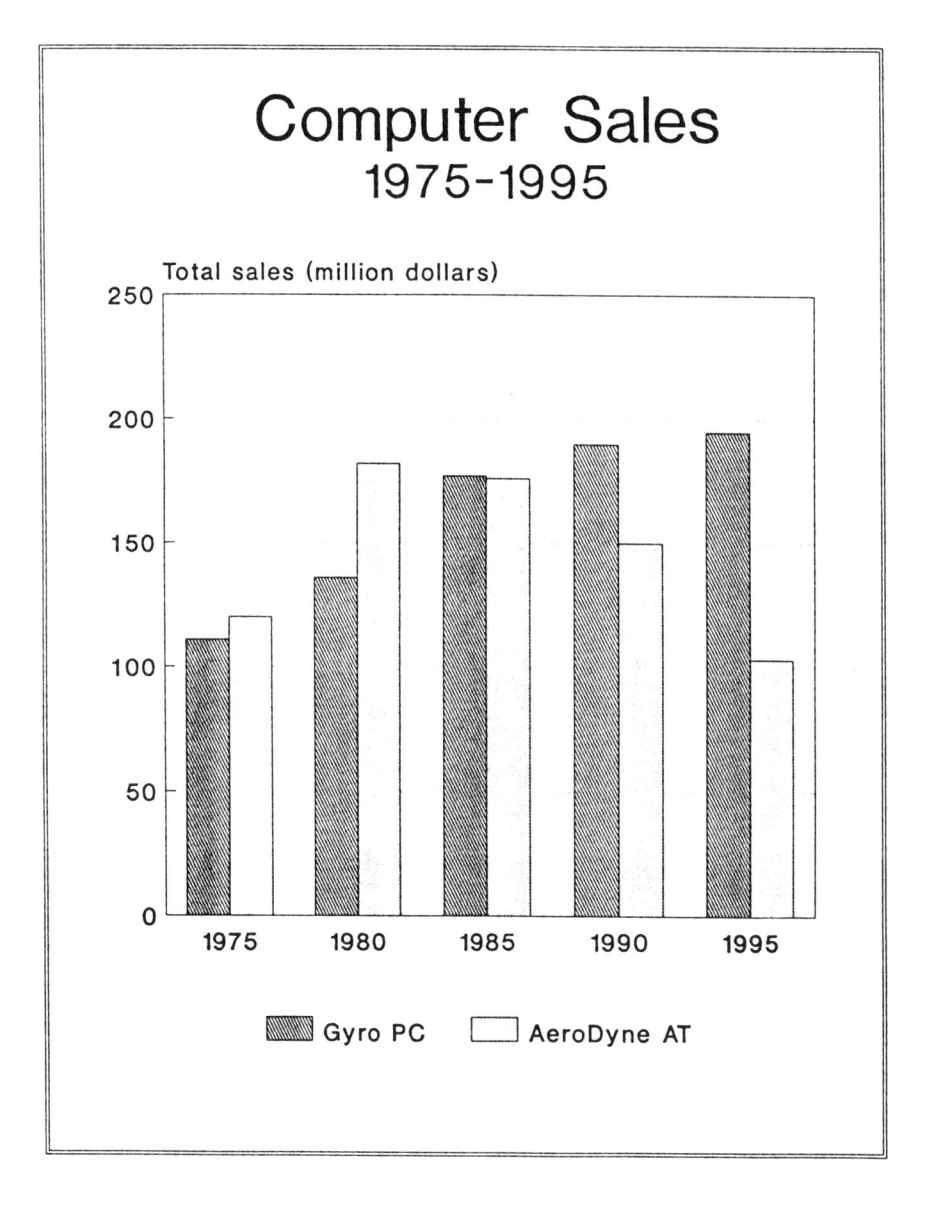

Figure 22-7. The same data from Figure 22-6 in the form of a bar chart.

TEXT CHARTS

Every presentation graphics program has the capability of producing text only charts that can be shown on the screen or used as a master for an overhead transparency. Most of these programs allow the use of different typestyles in almost unlimited sizes. Text may also be incorporated into drawn charts and graphs. *Bullets* can also be incorporated into these charts.

PAGE ORIENTATION

Page orientation may be either portrait or landscape. *Portrait orientation* means that standard paper is viewed as 8.5 inches wide and 11 inches high. *Landscape orientation* means that standard paper is 11 inches wide and 8.5 inches high. These programs allow the user to switch back and forth between the two orientations.

DRAWING

Most presentation graphics packages have drawing capabilities similar to the drawing programs mentioned earlier. The user can draw and place geometric shapes, lines, and text. This feature is also used to edit imported line art and/or photographs.

IMPORTING GRAPHICS FILES

Most of these programs include a fairly extensive selection of line art files. These can be imported using the drawing screen. Many also can import files saved in one or more of the standard graphics formats.

IMPORTING DATA FILES

Data files usually contain numerical information used to create pie, bar, or line graphs. The data may be imported from an ASCII file, a spreadsheet file, or a database file. Once the data file is imported, the user

and make multiple comparisons. Bar and line chart options include colors, patterns, spacing (for bar charts), labels for series, and labels for the x axis and y axis.

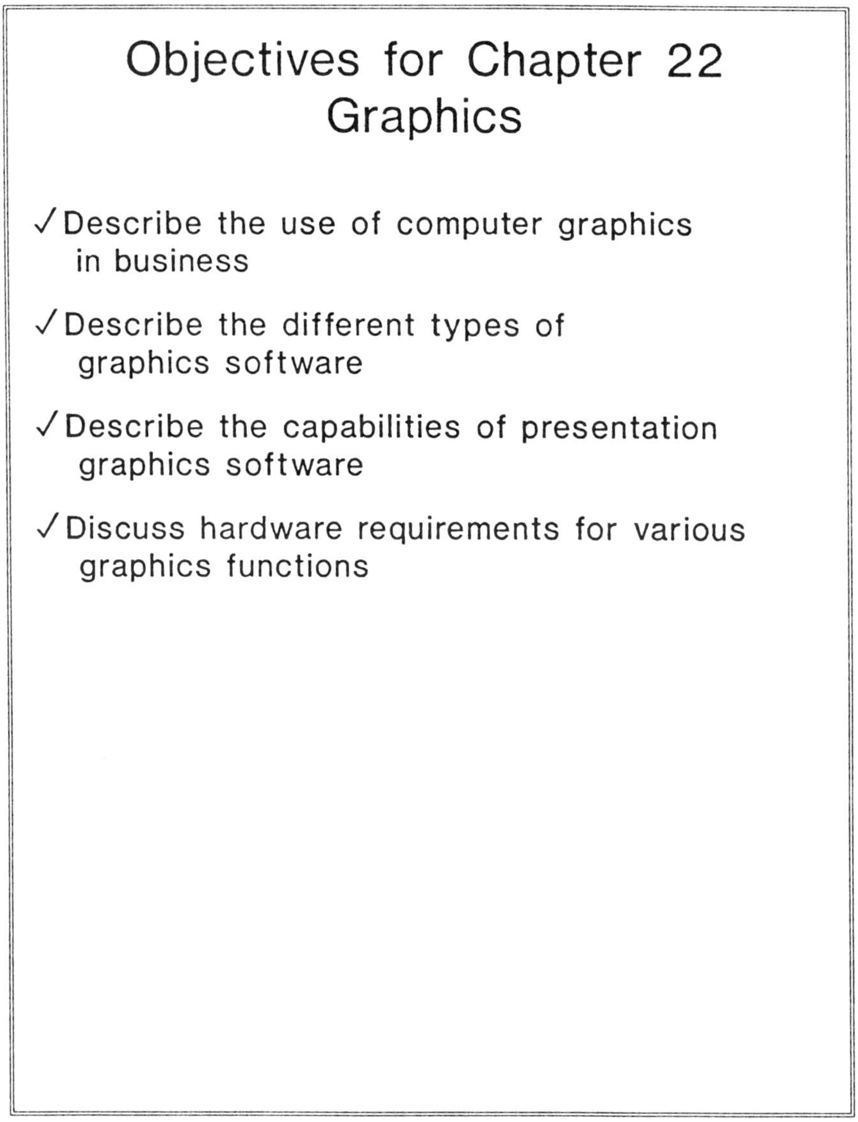

> **Objectives for Chapter 22
> Graphics**
>
> ✓ Describe the use of computer graphics in business
>
> ✓ Describe the different types of graphics software
>
> ✓ Describe the capabilities of presentation graphics software
>
> ✓ Discuss hardware requirements for various graphics functions

Figure 22-8. An example of a chart with bullet statements.

selects the type of graph and options and the chart is complete. Because the data is not re-entered, many possible mistakes are eliminated.

SCREENSHOW BUILDING

When a user has completed a file, he or she can save the file to a disk.

When this is done, the user has the option of adding it to the current screenshow. A *screenshow* consists of a group of chart files on a disk that have been specified for presentation as a group. Screenshows can be edited to add or delete files from the lineup. The user can customize the screenshow by specifying how one chart replaces the other on the screen and how the user indicates that it is time for the next chart to appear. Replacement of chart options include scrolling, replacing, fading and wiping. The user can select a timed presentation, where one chart automatically replaces the next after a specified time, or a key can be pressed to make the next chart appear.

OUTPUT CHOICES

One of the major features of presentation graphics programs is the different possible outputs that can be generated. Output may go to the screen or other viewing device, be sent to a printer, be sent to a plotter, recorded in a file for processing into a 35mm slide by a film recorder, or saved in a standard graphics format for use by another program.

Screen Viewing

In addition to viewing on the regular monitor, the computer can be hooked up to a projector which projects the images on a wall screen, or to a viewer that is placed on an overhead projector for viewing on a wall screen. The screenshow feature is often used in one of these ways.

Printing

Presentation graphics packages support a wide variety of dot matrix, laser, and inkjet printers. The resulting documents may be used in a portfolio or other document, or they may be used to make overhead transparencies. Most of the packages will allow the user to select the print quality (draft for first looks at a printout, high-quality for finished products). Of course, laser or inkjet printers with color printing capability are supported by these packages as well.

Plotting

A plotter is an output device that makes use of colored pens to draw

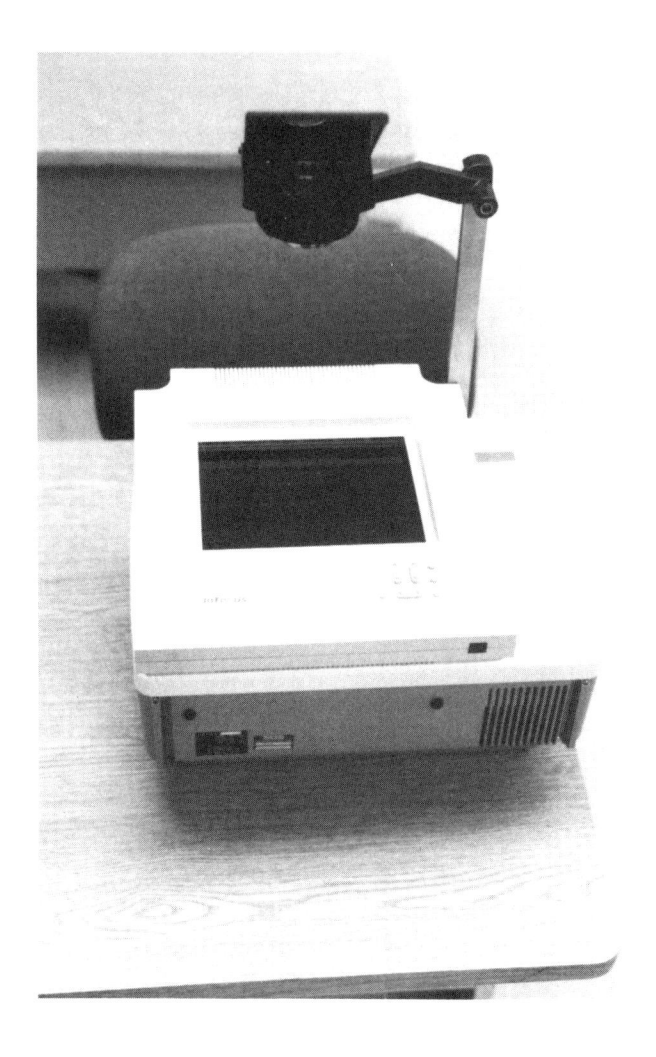

Figure 22-9. This viewer allows the user to send the output that normally goes to the monitor to an overhead projector for displaying on a wall screen.

graphs and line drawings on paper or transparency film. Most presentation graphics packages have drivers to support plotting of pie, bar, and line graphs. The result is a high-quality document that might be included in a report or shown as an overhead transparency.

35mm Slides

Most presentation graphics packages give users the option of saving files into a format to be made into 35mm slides by a film recorder. The slides may be produced in-house or sent to a custom slide service bureau such as Autographix or MagiCorp. These companies will develop the slide from a floppy disk.

HARDWARE REQUIREMENTS FOR COMPUTER GRAPHICS

To get the maximum benefit from a graphics program, a user must have a computer system that supports the software adequately. The most popular personal computer operating system for production of graphics has been the Macintosh system. If a business plans to produce almost exclusively graphics, a Macintosh is recommended. With IBM-compatible computers, the Microsoft Windows or OS/2 environments are generally preferred over DOS.

Graphics files that are bitmapped require a large amount of storage space. If the business plans to do a lot of graphics work, a hard disk in the 200MB range is recommended. A CD-ROM disk is also a good investment,

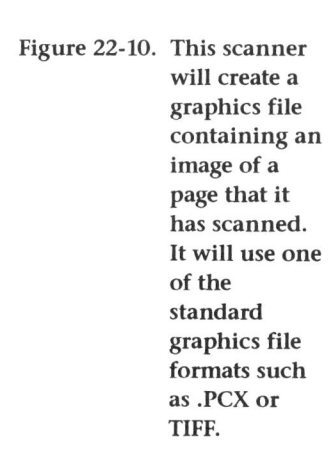

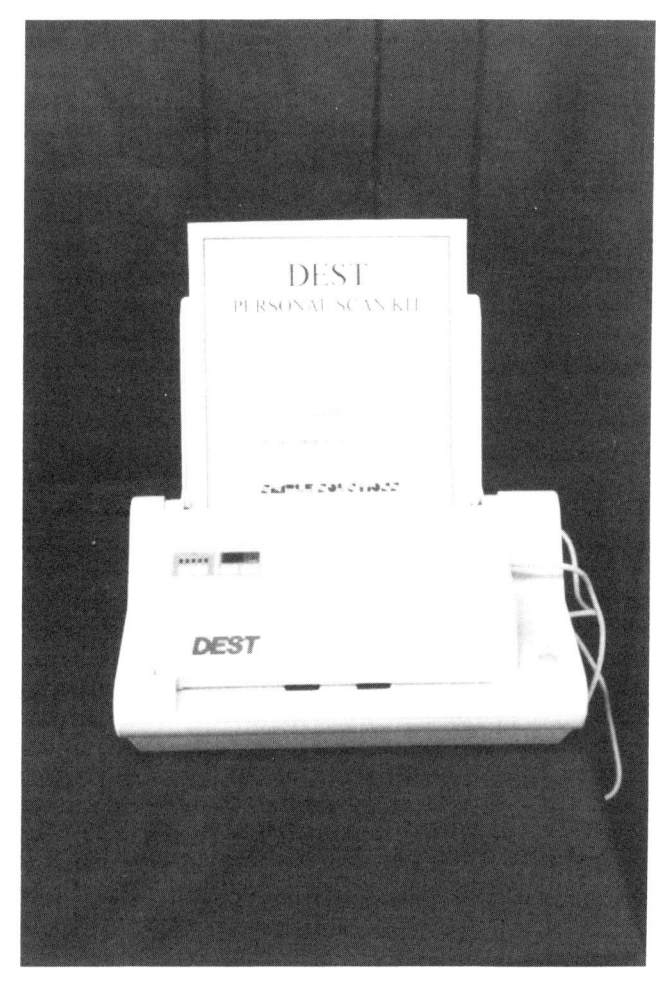

Figure 22-10. This scanner will create a graphics file containing an image of a page that it has scanned. It will use one of the standard graphics file formats such as .PCX or TIFF.

as many packages are starting to provide extensive graphics libraries on CD-ROM. For the same reason, the computer for graphics programs should have adequate memory—at least 4MB of RAM.

A scanner can save a lot of time by scanning graphics images (and text) from paper copies without having to redraw or retype them. The user should be sure that the scanning software will save scanned files in a standard graphics format that can be used with other applications.

CAD users will probably want to invest in a math coprocessor. The CAD software uses *floating-point calculations* for production of the designs and different views shown. A math coprocessor will cut the processing time almost in half for these programs.

A monitor for graphics production must be a color monitor. The Super-VGA standard is preferred because it will produce the sharpest images. Another consideration is the size of the monitor; at least a 14-inch monitor is recommended. Desktop publishing users often select a 17-inch monitor, so the entire page of a document can be seen in WYSIWYG format without scrolling.

The capability of the output device is often a limiting factor in the

Figure 22-11. A portable computer running Harvard Graphics. This computer is a wise choice for users who need to travel to different places and make presentations on a regular basis.

production of quality graphics. A laser printer will serve as a good general-purpose printer and also produce good graphics. Users should be aware of the memory available when purchasing a laser printer for graphics. Most laser printers come with 1MB of memory standard, not enough to produce high quality graphics using the presentation graphics packages. A minimum of 2MB of printer memory is recommended. Plotters and color printers should be considered by users who expect to produce a large quantity of graphics.

A film recorder is usually not necessary for most businesses. Slide services will produce high-quality slides, usually in less than two weeks.

One option that users may want to consider is buying a portable or notebook computer. For those users who have to travel and make presentations, a notebook computer combined with a viewer for an overhead projector works well. Newer notebook computers come with VGA monitors, docking stations with extra storage space, and CRT monitors.

CHAPTER QUESTIONS

1. What are presentation graphics?
2. Briefly describe the five major types of charts/graphs.
3. What is a standard graphics file format?
4. Briefly describe the five major types of graphics software.
5. Briefly describe five features of presentation graphics software.
6. What operating system(s) are considered the best for users who will be doing primarily graphics work?
7. Why do CAD programs run faster with a math coprocessor?
8. What type of user would benefit most from buying a portable computer?

Chapter 7

COMMUNICATIONS/NETWORKING

The combination of computer technology and communications technology has created an environment in which almost anyone can get information from almost anywhere. A farmer can check the current market prices of commodities using a computer and a database network such as AgriData. A tractor dealer can order replacement parts by punching a few buttons on a computer. A manufacturer in the United States can send a question to a supplier in Australia and have the answer back in seconds. Never before has the world been as connected as it is in the 1990s.

This chapter discusses some of the concepts that are important to computer communications and networking. Types of networks, requirements, operating systems, and communications software are presented.

OBJECTIVES

1. Define networking and data communications.
2. Describe components of a data communications system.
3. Describe the functions of networks.
4. Identify the types of networks.
5. Describe the configurations of local area networks.
6. Identify UNIX operating system functions and commands.

USING THE COMPUTER AS A COMMUNICATIONS NETWORKING TOOL

Data communications is the transfer of information from one computer to another over a communications channel. The primary method is by telephone lines, although use of other methods is increasing.

A computer *network* is a group of computers or terminals that are connected for the purpose of communicating. The primary purpose of networking is to share information—data files, software, messages—without having to physically carry disks or play "telephone tag" and to share hardware, such as storage space, printers, or microprocessors.

A single communications line directly connecting one computer to another is called a point-to-point line. If more than one computer is connected, it is referred to as a multi-drop line.

COMPONENTS OF A DATA COMMUNICATIONS SYSTEM

A data communications system has four major components: the computers, the transmission channel, the communications hardware, and the communications software. Computers have been discussed in previous chapters. The other three components are described below.

COMMUNICATIONS CHANNEL

The *communications channel*, also called a link, is the means or path by which the data gets from one computer to another. The three most

---

TERMS

analog signals	electronic mail	node
baud rate	Ethernet	protocol
bits per second (bps)	fiber optic cable	ring network
bus network	Internet	server
coaxial cable	local area network (LAN)	star network
communications channel	modem	terminal emulation
data communications	network	twisted pair wire
digital signals		

common types of communications channels are twisted pair wire, coaxial cable, and fiber optic cable.

Twisted Pair Wire

Twisted pair wire is the type of wire used for telephone lines. Telephone lines are the most common communications channel used in data communications systems. This method of communicating is inexpensive, but data can be adversely affected by electrical interference, called noise.

Coaxial Cable

Coaxial cable is a high quality wire commonly used in cable television systems. This cable is not susceptible to interference and can transmit data faster than a twisted pair wire.

Fiber Optic Cable

Fiber optic cable is made of thin strands of glass, not copper wire like twisted pair wire and coaxial cable. The cable is light, inexpensive and does not have to be protected from moisture because it does not rust. Data transmission is very fast with no interference. Most data communications systems developed will use fiber optics. It is already replacing twisted pair wire in many locations.

Other Communications Channels

In addition to the three most common channels, three others are sometimes used in data communications systems. Microwave transmission will transmit voice and data signals over radio waves, but only in line of sight transmission. Satellite transmissions are used to transmit data in some instances, usually when long distances are involved. Wireless transmission uses radio waves or light beams. Some wireless transmissions work like a cellular telephone, while others use existing electrical wiring to transfer the radio signals.

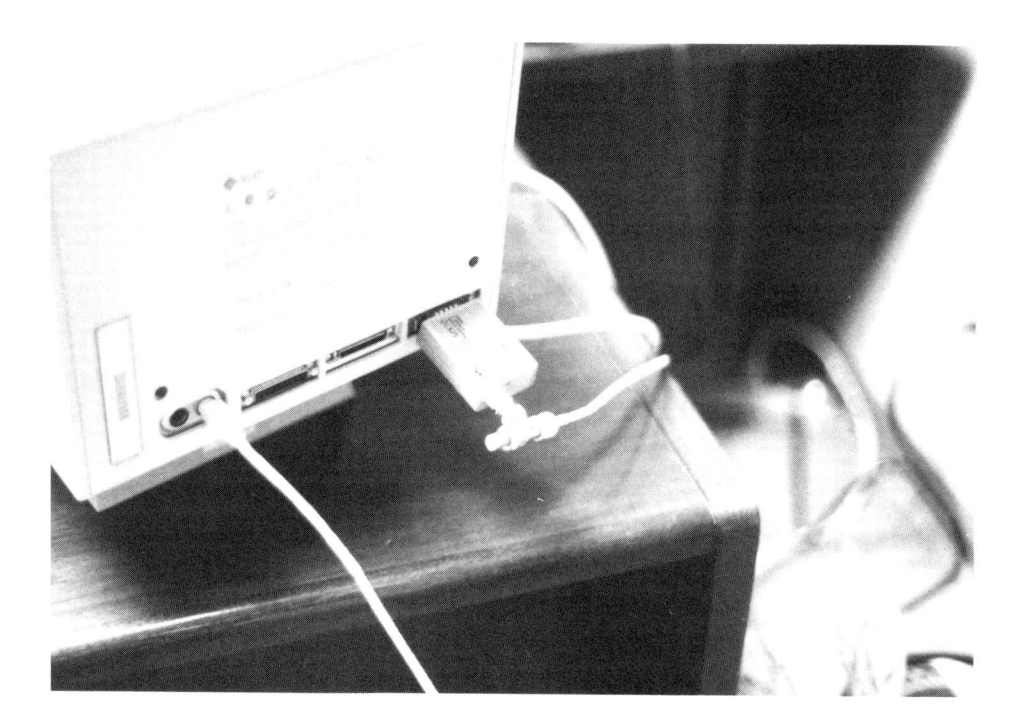

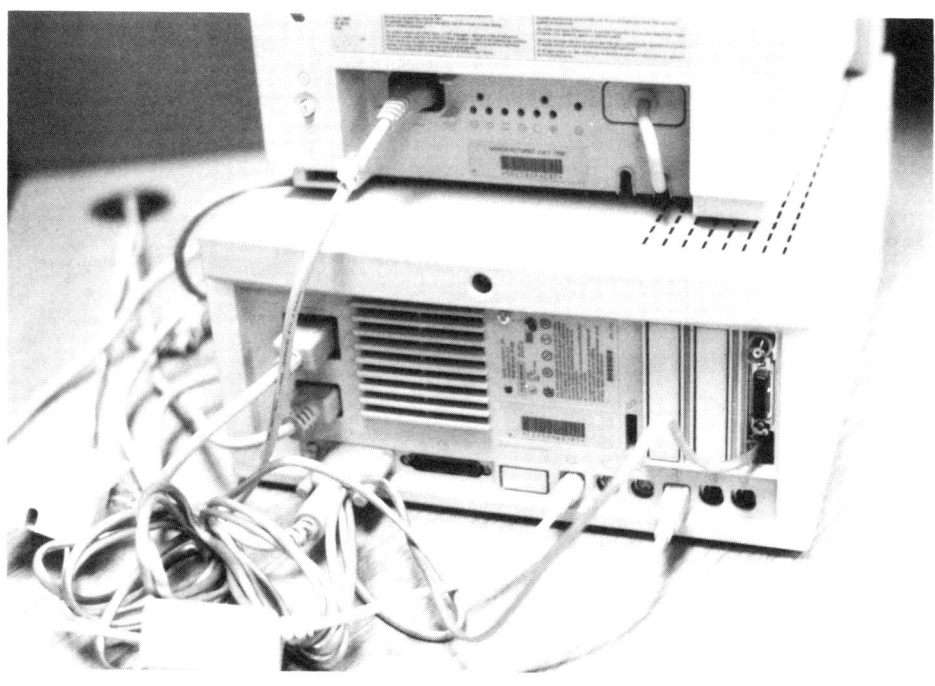

Figure 23-1. These two machines are both connected to a network. Above is a terminal connected to a mainframe with coaxial cable. Below is a Macintosh which is connected to a LAN by use of the twisted pair telephone wire (last connection on the right).

COMMUNICATIONS HARDWARE

Within short distances, up to approximately 1,000 feet, computers can be connected by wires from serial port to serial port. For further distances, special equipment must be used to convert the computer signal to a communications signal and then convert it back to a computer signal when it reaches its destination. A *modem* is the device used to do this.

The name modem is derived from the terms *mod*ulate-*dem*odulate. A computer communicates by sending *digital signals*, distinct electrical impulses. A communications channel, such as twisted pair wire or fiber-optics cable, transmits *analog signals*, electrical waves. The modem modulates (converts the digital signal to an analog signal) data going out of the computer. It demodulates (converts an analog signal to a digital signal) data coming into a computer. For two computers to communicate at long distances, both must have a modem.

Modems can be described in several ways: type of hookup to the computer, data transmission rate, and method of transmission.

Figure 23-2. An external modem.

Modem Hookups

Modems are classified as either internal or external. Internal modems usually occupy an expansion slot in the computer and the communications lines plug directly into the back of the computer. External modems connect to the serial port of the computer. The advantage of external modems is that they can be disconnected easily and transferred for use with another computer. The disadvantage is that they take up room on the desktop and may get knocked off and damaged.

Data Transmission Rate

The data transmission rate of modems is expressed in either *bits per second (bps)* or *baud rate*. Bits per second is the number of bits per second that can be transmitted. Common speeds are 1200 and 2400 bps. The baud rate is the number of times the signal changes per second. One or more bits may be transmitted each time the signal changes. This means that the actual bits per second will be higher than the baud rate. A modem with a baud rate of 4800 may have a bps of 9600 or more.

Method of Transmission

The three methods of transmission of data are simplex, half-duplex, and full duplex. Simplex refers to transmission in only one direction. It is almost never used. Half duplex means that transmissions can occur in both directions, but not at the same time. Full duplex means that transmissions can occur in both directions at the same time.

COMMUNICATIONS SOFTWARE

Communications software provides the link between the user and the data being transferred. The three primary functions are telephone dialing, file transfer, and terminal emulation.

Telephone dialing simply means that the software is dialing the telephone number of the computer to be connected. Most software systems provide a dialing directory which the user can customize with commonly used numbers.

File transfer is the ability of the software to send and receive files over the communications channel. The software uses a standard *protocol* so both

COMMUNICATIONS/NETWORKING

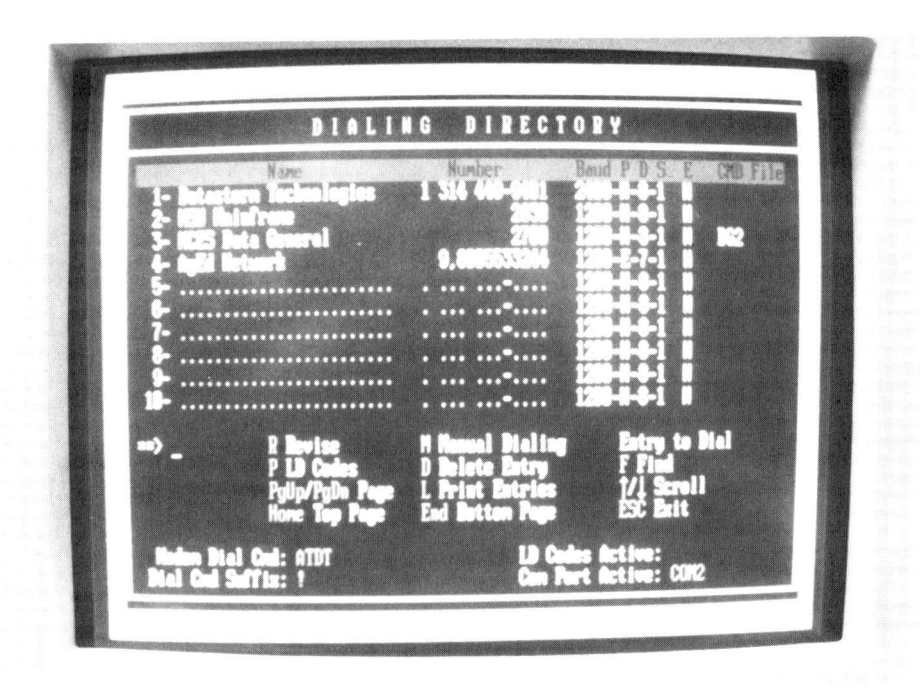

Figure 23-3. A dialing directory from the ProComm communications program.

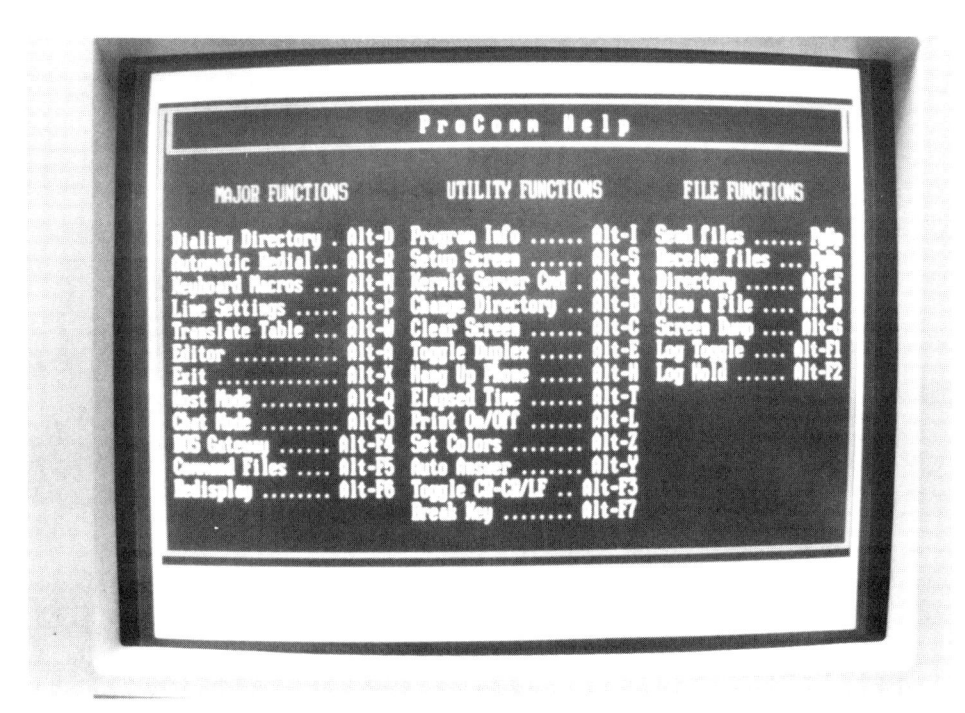

Figure 23-4. The help screen in ProComm. This feature makes the software package user friendly.

computers can interpret the file being transferred. Standard protocols for file transfer include Kermit, Xmodem, Ymodem, Zmodem and several variations of each.

When connected with some network computers, the personal computer may have to emulate a specific type of terminal in order to communicate. *Terminal emulation* is provided by the communications software. Some popular communications programs include ProComm, Telix, Smartcom, Qmodem, HyperAccess and Crosstalk.

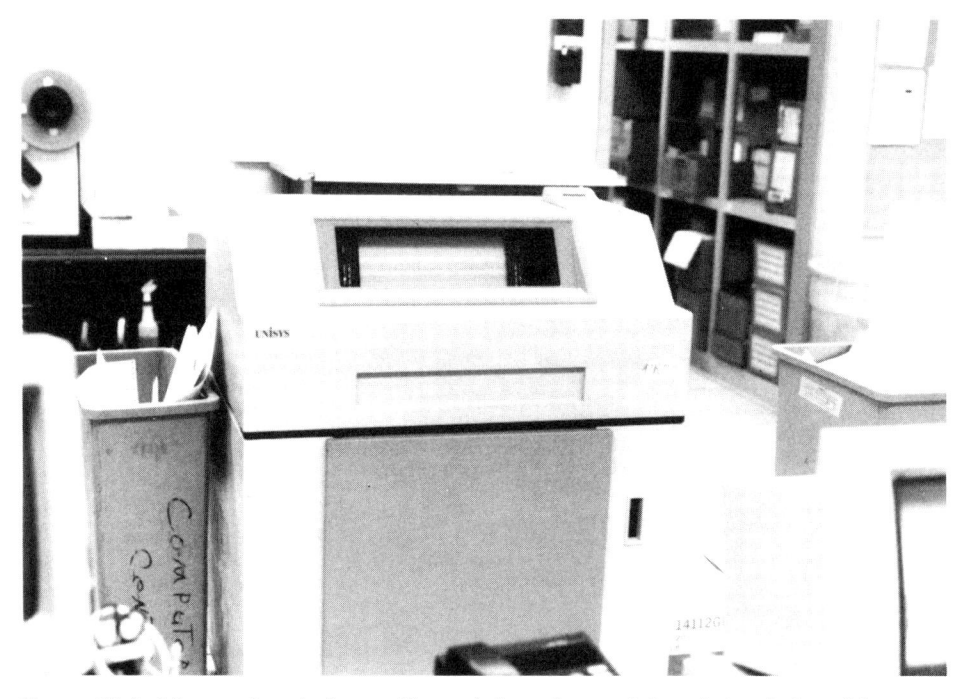

Figure 23-5. Many networks have a line printer, a heavy duty printer designed to handle the large printing loads of a network.

TYPES OF NETWORKS

Any time two or more computers are connected, they are considered to be networked. The two commonly used descriptions of networks are local area networks (LAN) and wide area networks (WAN).

LOCAL AREA NETWORKS (LAN)

A *LAN* is a network of computers located in close proximity to each other. These networks usually are located in an office, building, or group of buildings. LANs are privately owned networks, usually with a series of personal computers connected via a communications channel, such as an *Ethernet* cable or telephone cord.

The functions of a LAN include sharing data files, sharing applications software, messaging (electronic mail), and sharing resources. The resources might include a network printer, usually a laser or a line printer, or a gateway. The gateway is a means of communicating with a different type of network outside the LAN.

Some LANs have a computer designated as a *server*. The server may be one of the personal computers or a minicomputer. A file server is used mostly for handling files and routing requests. A client server performs most of the processing for the computers on the network. The server must be operating for the system to work. It is important for the system administrator to keep a current backup of the disk storage on the server so system failure does not cause users to lose data.

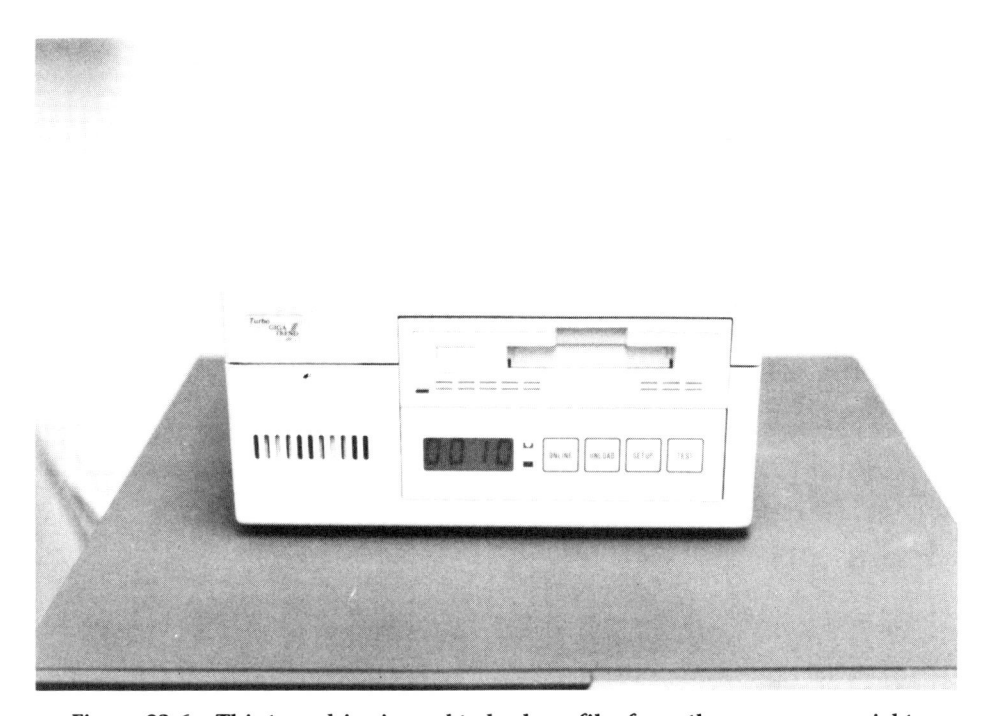

Figure 23-6. This tape drive is used to back up files from the server every night.

WIDE AREA NETWORKS (WAN)

A *WAN* is different from a LAN in two ways. It is spread out more geographically and it uses telephone lines or satellite transmission to communicate between computers.

WANs include commercial database networks that provide a variety of services for their users: CompuServe, Prodigy, America Online, GEnie, and Delphi are large commercial networks. These networks serve as information clearing houses and provide *electronic mail* services, catalog shopping, weather updates, and much more information. Users pay a monthly subscription fee and may have to pay more for special services.

Agribusinesses may benefit from subscribing to the AgriData Network, which provides agricultural market updates, agricultural news, and other services. Other on-line sources of agricultural information include the National Agricultural Library, USDA, and Doane's Agricultural Computing network.

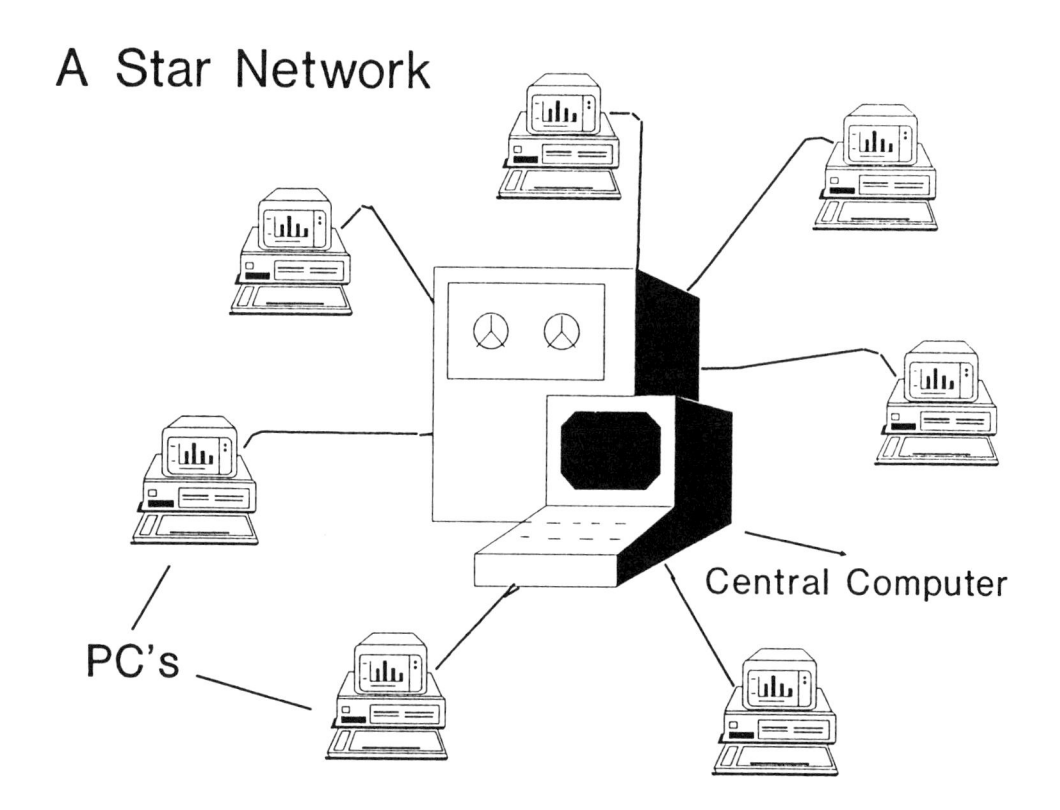

Figure 23-7. Diagram of a star network.

CONFIGURATIONS OF NETWORKS

Networks may be configured in a number of different ways. A *star network* has a central computer with various personal computers and/or terminals attached. This is an efficient configuration, but problems with the central computer shut down the entire system.

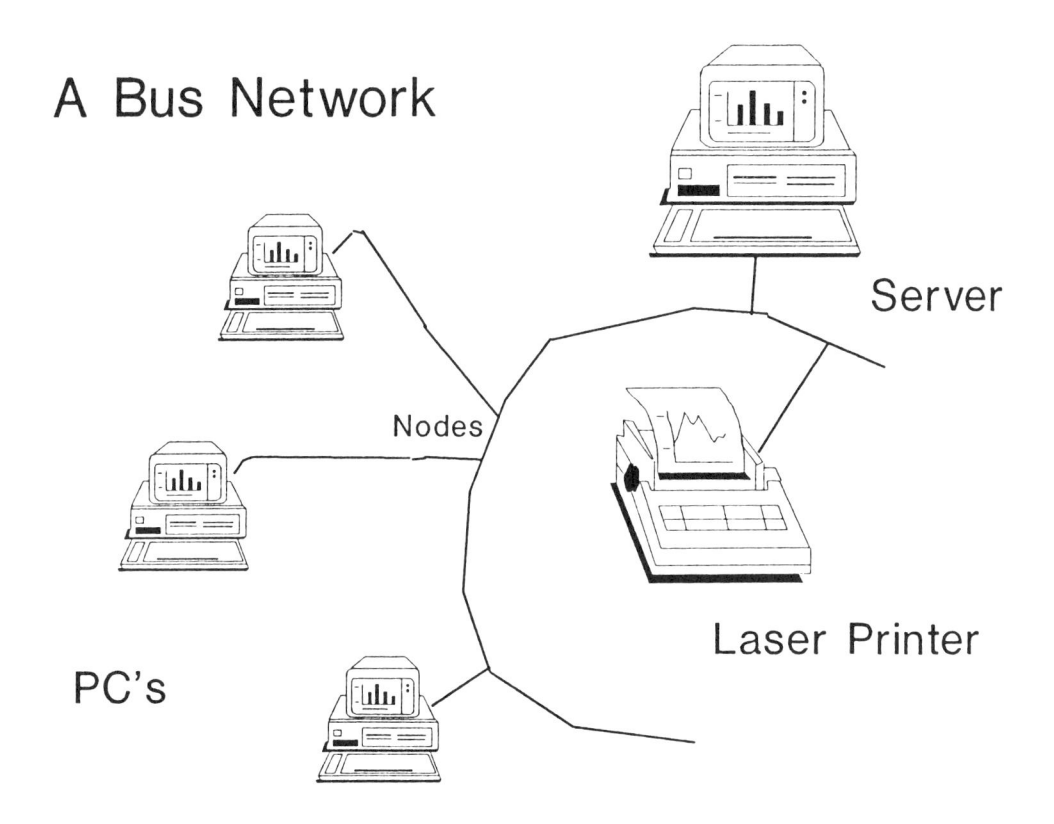

Figure 23-8. Diagram of a bus network.

A *bus network* has several computers connected on a single communications line. The duties of network management may be shared or one machine may be used as a server. Peripherals can be attached or removed without hurting the system. When one computer goes down, the rest are usually not affected. The majority of LANs use a bus configuration with an Ethernet networking system and high speed cable.

A *ring network* does not use a central computer or server. Data passes

A Ring Network

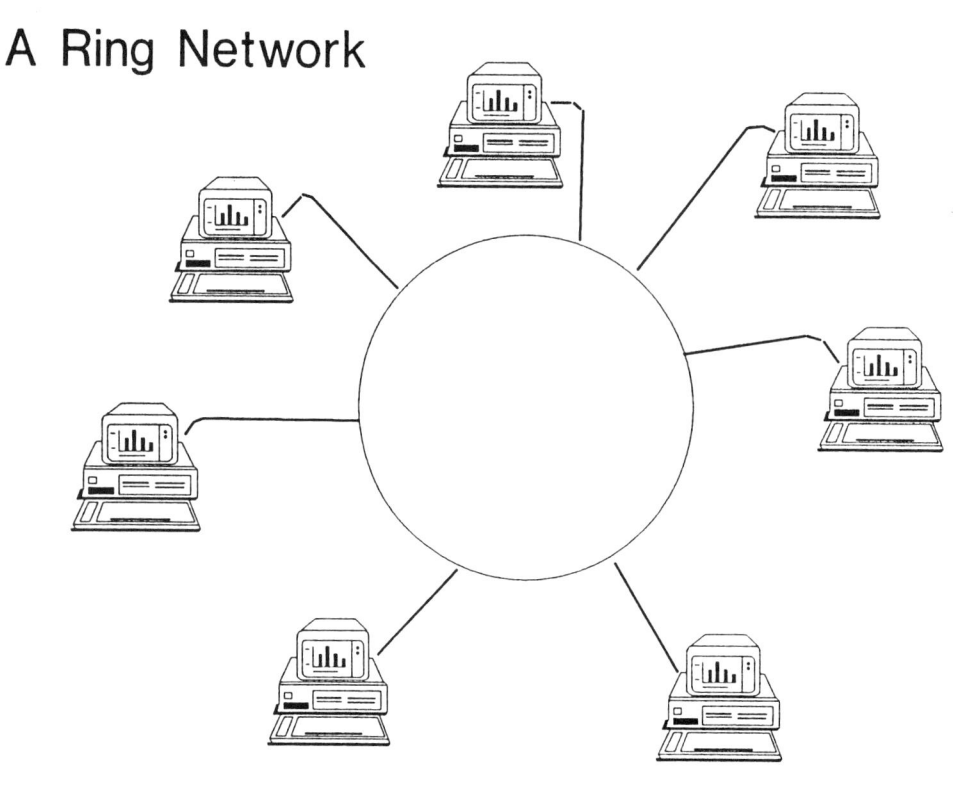

Figure 23-9. Diagram of a ring network.

around the ring through each *node* (computer) on the network. If one computer breaks down, the entire network is down.

THE UNIX OPERATING SYSTEM

As mentioned in Chapter 17, UNIX is a multitasking, multiuser operating system. As such, it provides password protection for users' files. These characteristics make UNIX a very popular operating system for networks.

When using UNIX, users should remember that it is case-sensitive, unlike DOS. UNIX commands are entered in lower case. Table 23-1 contains a summary of UNIX commands and the similar DOS commands as quick reference.

UNIX has three significant capabilities that make it popular for data communications: electronic mail, file transfer protocol, and telnet.

ELECTRONIC MAIL

The electronic mail function in UNIX allows the user to send messages to all other users of the network. If the host computer is connected to other computer networks, such as the *Internet* or Bitnet, then the electronic mail function in UNIX will allow the user access to these networks as well. Each user is assigned a network address, usually the same as their login code.

Table 23-1
A COMPARISON OF UNIX AND MS-DOS COMMANDS

Function	UNIX Command	MS-DOS Command
Directory listing	ls	DIR
Copy a file	cp	COPY
Delete a file	rm	DEL
Change directory	chdir	CHDIR
Create a subdirectory	mkdir	MKDIR
Remove a subdirectory	rmdir	RMDIR
Print a filel	pr	PRINT
Display a file on screen	cat, more	TYPE, MORE
Online help	man	HELP
Send and receive mail	mail	
Obtain info about other users	finger	
Chat with other user	talk	
Write a message to a user	write	
List active users	who	
Remote terminal login	telnet	
Remote file transfer	ftp	

A user's Internet address is his or her network address, followed by an @ symbol, followed by the address of the host computer. The host computer address is divided into three parts: the computer name, the location, and a suffix indicating the type of institution, all separated by a period. For example, John Doe at AT&T may have the following address:

jdoe@computer1.att.com

The com suffix means that he is at a company. An edu suffix refers to an educational institution and a gov suffix refers to a government agency.

FILE TRANSFER PROTOCOL (FTP)

File transfer protocol is the means by which UNIX allows users to transfer files from computers at which they have a password. Anonymous ftp allows the user to transfer files which have been left open by the person at the remote site.

Anonymous ftp is used by accessing the remote computer using the ftp command. The user types anonymous at the login and his or her full address at the password prompt. Only certain directories and files on the remote computer system will be available.

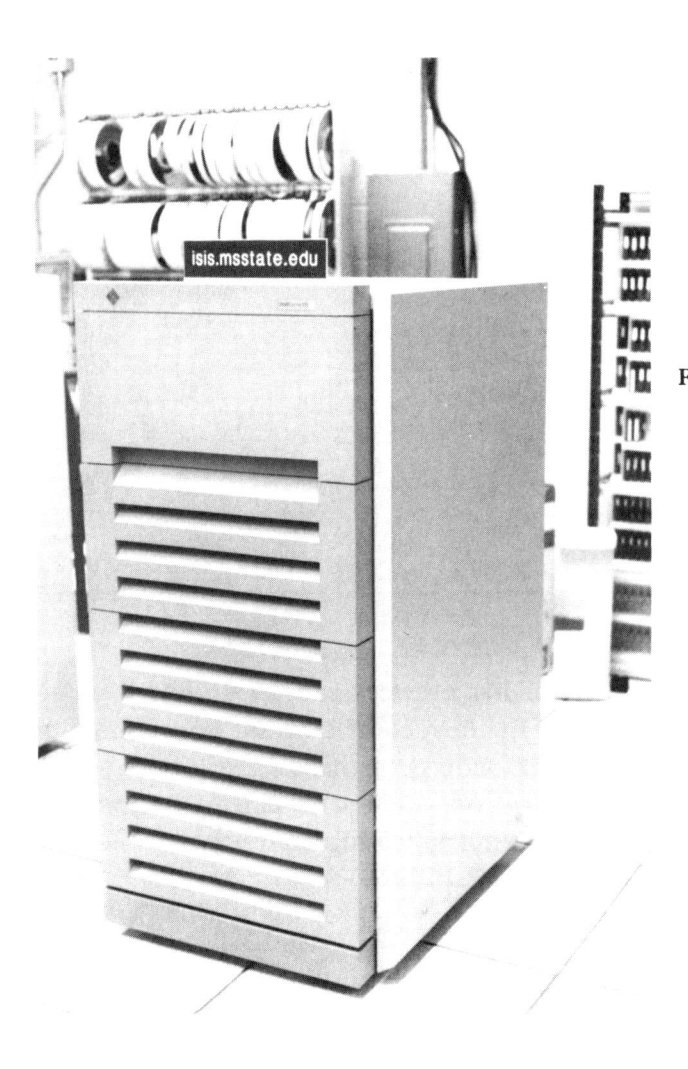

Figure 23-10. This Sun SparcServer functions as the central computer in a network at Mississippi State University. Note that it is connected to the Internet—the host address is isis.msstate.edu.

TELNET

The telnet command allows the user to login to a remote computer using a different operating system or a different version of UNIX. Some systems allow users to login and access bulletin boards, etc. Many educational and extension networks allow remote users to login to their systems and share or access information.

CHAPTER QUESTIONS

1. What is data communications?
2. What is networking?
3. What are the four components of a data communications system? Briefly describe each.
4. What does a modem do?
5. Which type of communications channel is the best?
6. What are the three ways of classifying modems? Briefly describe each.
7. What is communications software? What functions does it perform?
8. What are the two primary types of networks?
9. What are the three major configurations of networks?
10. Briefly describe the three major features of UNIX as a communications operating system.

Glossary

Accounting software—ready to use recording system based on general accounting practices
Address—the relative or actual location of data
Analog signal—electrical waves
Application—a computer program used for a particular kind of work, such as word processing or database management; synonymous with computer program
Artificial intelligence—simulated ability to acquire and retain knowledge
ASCII files—files organized using characters represented by American Standard Code for Information Interchange
AUTOEXEC.BAT—a user created program containing a list of commands which are executed at setup
Backup—to make a spare copy of a file or disk ensuring that the information won't be lost if the original is lost or damaged
Bar graph—a graph that shows numeric data as a set of evenly spaced bars
Baud rate—the transfer rate of data over a serial interface
Binary logic—binary, meaning two, is the number of distinctions the computer can make on the signals it receives
Bit—short for binary digit, one signal represents open or closed (off or on)
Bitmapped font—each character is described as a pattern of dots in a specific point size
Bitmapping—the process of using pixels to generate a graphic of an image as a digital image
Bits per second (bps)—an expression of data transmission rate
Blocking—marking a section of text
Boolean "AND/OR" comparisons—"AND" means that the search will include every unit with a certain value in one field and a certain value in another, "OR" means that the search will include every unit with a certain value in one field or a certain value in another field
Booted—the procedure of what the computer does when started up
Buffer—part of RAM used for temporary storage of data for input or output functions
Bullets—graphic symbols variable in size and used to highlight text
Business graphics—using graphics to convey information to clients, management, shareholders, etc.
Bus network—several computers connected on a single communications line
Byte—collectively eight bits with 256 possible on or off combinations used to represent all of the characters
Cathode ray tube (CRT)—the most common type of monitor for personal computers
Cell—a unit of a spreadsheet formed by the intersection of a column and row that stores data
Cell address—the column and row label intersection
Characters per inch (cpi)—the number of vertical characters in an inch for a printer
Coaxial cable—a connecting cable consisting of two insulating layers and two conductors
Code—command or function, as opposed to text
Command files—files used by the operating system and utilities
Commercial software—software copyrighted by the owner and illegal to copy, other than to make backup copies for personal use
Communications channel—a data communication path that allows data transmission to travel from its origin to its destination
Companion programs—associated programs required for certain functions
Computer aided design (CAD)—software used to create original technical drawings
Concordance file—a list of important words
CONFIG.SYS—a file created when DOS is installed specifying buffers, location of files, etc.

which can be user modified
Conventional memory—the first 640K of RAM memory in a DOS based system
Cursor—the pointer on the computer screen that indicates current location
Cursor control keys—arrow keys and others used to move the cursor around the computer screen
Data—facts or figures from which conclusions can be inferred, information
Data communications—the transfer of information from one computer to another over a communications channel
Database—an organized collection of related information
Data files—files created within an application program
Daisy wheel printer—an impact printer based on technology available in many typewriters
Device drivers—the software that interfaces a particular peripheral to the microprocessor
Dialog box—a window which appears temporarily to request information
Digital signals—distinct electrical impulses
Directory—a computer index of files, programs, or other directories on a disk
Disk drive--a device used for recording and retrieving information on disks
Document—grouping of words, sentences, paragraphs, and pages
Documentation—supporting references
Document layout—positioning text or graphics in certain ways in a document
Dot matrix printer—prints characters composed of dots by pressing wires against an ink ribbon and onto the paper
Dot pitch—the physical size of the pixel on a monitor
Editing—making changes in an existing document
Electronic mail—the ability to send personalized messages via personal computers from one to another
Ethernet—a type of connection for a LAN communication channel
Extension—up to three characters in length preceded by a period (.) following the filename
External commands—commands not part of the operating system kernel and each has its own filename as a part of DOS files
Fiber optic cable—cable made of thin strands of glass which allows light to pass through
Field—a defined area which contains one kind of information, a labeled column
File allocation table(FAT)—the address location of a file on disk recorded in table format
File—a collection of information that has been given a name and stored on a disk
File management—the techniques of organizing and maintaining data
Filename—the name assigned to a computer file
Film recorder—an output device to make 35mm slides from an exported graphics file
Fixed-pitch font—uses the same space on a page for each character
Floating-point calculations—a system of arithmetic involving having the numbers expressed in scientific notation for calculations, e.g. .0005 becomes 5×10^{-4}
Floppy disk—a disk that can be inserted in and removed from a floppy disk drive; usually 3½" or 5¼"
Font—a complete alphabet of a typestyle in one size
Font renderer—a scheme to interpret the mathematical formula of a scalable font in a printable character
Footers—printed at the bottom of each page
Formula—a mathematical expression or calculation entered into the cell of a worksheet
FORTRAN—an acronym for Formula Translation, the first computer programming language to use English commands
Freeware—freely-distributed software that has been placed in the public domain
Function keys—10 to 12 keys on a computer keyboard that perform specific actions in application programs
Gigabyte (GB)—1,024MB or 2^{30}
Grammar checker—software designed to accomplish the editing and proofreading function

GLOSSARY

Graphical user interface (GUI)—a computer interface characterized by the use of a mouse to access its pull-down menus and icons
Graphics--the ability of the computer to display images
Graphics drivers—a program that translates operating system requests into a format that is recognizable by specific hardware
Hard disk--a magnetic storage medium, usually much larger than floppy disks, used to store files and software
Hardware--the equipment that makes up a computer; such as the keyboard, printer, monitor, mouse, disk drive, etc.
Headers—printed at the top of each page
IBM compatible computer—IBM clones, computers that use the DOS operating system
Icons—symbols representing computer commands
Inkjet printer—a printer using a spray nozzle to spray ink on the paper to produce a character or graphic
Input device—any computer peripheral that sends information to the computer's microprocessor
Integrated software—several software applications packaged together and sold at one price
Internal commands—those commands most commonly used by users and application programs which are loaded in RAM when the system is booted
Internet—a computer connected to other networks allowing the user access to these networks as well
Kernel—the central core of information of the operating system resident in RAM memory
Kilobyte (K)—1,024 bytes
Label—descriptive text in a worksheet
Landscape orientation—printing across the page length
Laser printer--a computer printer using laser technology producing a page at a time
License fee--the money paid for the formal permission to do something
Line art—artwork composed of line drawings
Line graph—a graph that represents numeric data as a set of points along a line
Liquid crystal display (LCD)—images displayed by using a liquid crystal material between two pieces of glass
Local area network (LAN)—a series of personal computers connected via a communications channel which are located in close proximity to each other
Login—the process of signing on an on-line computer network
Logout—the process of signing off an on-line computer network
Machine language—the actual binary language the computer uses to function
Macro—one instruction which produces a series of action to be accomplished
Margins—top, bottom, left and right blank area of page
Math coprocessor—a chip in conjunction with the microprocessor which may double its processing speed with math-intensive software
Megabyte (MB or Meg)—1,048,576 bytes or 2^{20}
Memory—the part of computer storage directly accessible to the microprocessor; usually synonymous with random access memory (RAM)
Menu—a list of choices presented by a computer program
Merging documents—the combining of two or more separate documents
Microchip—highly miniaturized integrated electronic circuits on a silicon chip often smaller than a dime
Microprocessor—the integrated circuit "chip" component within the computer that directly executes instructions
Microsoft Windows—an operating environment used with MS-DOS that is a graphical user interface with multitasking capabilities
Millions of instructions per second (MIPS)—term used to express microprocessor speed
Modem—a communication device that enables a computer to transmit over a telephone line

Mouse—a small hand-held device used to control a pointer on the computer monitor
Multimedia—a combination of various media; such as sound, graphics, animation, and video
Multitasking—an operating system's ability to concurrently execute more than one application
Near letter quality—a mode of operation in a dot matrix printer which prints each line two times making it close to letter quality
Network--two or more computers connected by cables or other means and using software that enables them to share equipment and exchange information
Node—any network station
Numeric keypad—a separate area on the keyboard with the numeric keys arranged similar to a calculator keypad
On-line time—the actual amount of time a user is signed on a computer network
Operating system--the system program which controls the hardware and interfaces with user software programs
Output device—any computer peripheral that receives information from the computer's microprocessor
Page description languages—programming languages that describe the placement and appearance of text and graphics on a page
Parameter—value following a DOS command which tells the command where to perform its function
Peripherals--input or output devices that connect to the computer's microprocessor
Pie graph—a graph that compares parts to the whole
Pixel—the dots that can be illuminated on a monitor or printed creating a graphic or text
Point-and-click—a user interacting with a computer interface characterized by use of a mouse to access icons and pull-down menus
Portrait orientation—printing across the page width
Presentation graphics—includes the special effects of color, patters, illustrations, etc.
Program—a detailed set of instructions written by a computer programmer
Program file—a file used to operate an application program
Programmer—one who writes the detailed set of instructions for the computer to execute
Prompt—where the computer user gives commands by typing them in
Prompt area—display area for the command line and entry/status line for an electronic spreadsheet
Proportionally-space font—varies the space given to each letter based on the width of the letter
Protocol—a standard method for determining how and when to format and send data
Pull-down menu—a graphical user interface of commands or options which may be accessed with a mouse
Query-by-Example—a technique relying on names and key words with some national characteristics used by most of the popular programmable database managers
Random access memory(RAM)—memory used by applications to perform necessary tasks while the computer is turned on. When turned off, RAM is cleared
Range—a cell or contiguous group of cells
Read only memory (ROM)—memory which can be read, but not modified
Record—a collection of the information for all fields on a single subject
Relational database—a database with common data between fields and records
Report—a collection of individual records in printed form
Resolution—the number of dots that make up an image on a monitor or scanner
Ring network—does not use a central computer or server, data passes around the ring through each node
Read only memory (ROM)—memory which can be read, but not modified
Root directory—the base directory created when a disk is formatted
Scalable font—each character is described as a mathematical formula

GLOSSARY

Scale—ratio between the dimensions of a representation and those of the object
Screenshow—a group of chart files on disk that have been specified for presentation as a group
Scrolling—allows the computer user to view the information ahead and back, as well as sideways
Search and replace—finding an equal condition and changing the data
Semi-conductor—a substance used in transistors, rectifiers, etc.
Server—a computer that provides disk space, printers, or other services to computers in a network
Software--the set of computer instructions that make the hardware perform tasks
Software piracy—software given to others without the permission of the copyright holder
Spooling—temporary storage for information being output to a slower device
Spreadsheet—a grid usually made up of 256 columns and 8,192 rows
Standard graphics file format—established formats developed for graphics files
Star network—a central computer with various personal computers and terminals attached
Subdirectory—a group of files or lower child directories that contain related information
Switch—options as a part of a DOS command
System security—a procedure, usually through passwords, which limits access to computer applications
Template—a basic pattern for a spreadsheet
Terminal—the term used for a work station in a computer network
Terminal emulation—the imitation of all or part of one terminal by another so the mimicking device can accept the same data and perform the same functions as the actual device
Terminate-and-stay-resident (TSR)—application programs loaded into high memory at startup but not yet activated for use
Text features—commands that change the text from normal to text with special characteristics
Thesaurus—a list of synonyms and antonyms for a word
Time slices—period of time allocated by the operating system to each user
Transistor—a solid state, electronic device, composed of semiconductor material, that controls current flow without the use of a vacuum
Twisted pair wire—a pair of copper wires, insulated, and twisted together, telephone line
Undelete—the ability to recall deleted information stored in a buffer in RAM
User-supported software—software that can be freely copied and distributed, but the user is expected to register with the author and pay a fee for using the software
User-friendly—the ease with which interaction takes place
Value—a number entered into the cell of a worksheet
Virtual memory management—the operating system increases available memory by allocating a portion of it to hard disk
Virus—a term for anything undesired that attaches to or corrupts files or programs in a personal computer
Virus protection—those utilities that scan disks for the presence of a virus
Wide area network (WAN)—a series of personal computers spread out geographically and using telephone lines or satellite transmission to communicate between computers
Widow/orphan protection—elimination of the first line of a paragraph printed on the last line of a page or the   last line of a paragraph printed on the first line of a new page
Wildcard—the asterisk (*) and question mark (?) when using DOS commands
Window—a rectangular area on a computer screen in which an application or document is viewed
Worksheet—an electronic spreadsheet grid usually made up of 256 columns and 8,192 rows in table format

Index

Accounting software, 118-119
Binary logic, 12-13
Business graphics, 133-134
Communications/networking, 42, 151-165
    components, 152-153
    confifurations, 161-162
    definition of, 151-152
    hardware, 155-156
    software, 156-158
    types of, 158-160
    UNIX operating system, 162-165
Computer
    buying, 42-45
    defined, 2
    generations, 7-9
    history, 4-7, 10-11
    how computers work, 12-14
    memory, 13-14, 18, 50
        management, 78-79
        RAM, 13-14
        ROM, 13
    parts of, 14-36
        CD-ROM, 26-27
        expansion slots, 20-21
        floppy disk, 22-24
        hard disk, 24-25
        keyboard, 31-32
        microprocessor, 16-18, 49-50
        monitor, 28-31
        motherboard, 16-18
        parallel port, 27
        power supply, 15-16
        printer, 32-36
        serial port, 27
        tape drive, 26
    start up, 56-58
Database management, 40, 121-130
    concepts, 124
    definition of, 121-122
    features, 125-130
    functions, 122
    hardware, 124
    software, 122-124
Electronic mail, 163
Graphics, 121-123, 133-149
    definition of, 121-122
    features, 140-146
    hardware requirements, 147-149
    output choices, 145-146
    software, 135-140
        CAD programs, 137-139
        canned programs, 139
        other programs, 140
        printing/drawing programs, 136-137
        presentation packages, 139-140
Microsoft Windows, 54
Monitor
    Cathode Ray Tube (CRT), 28-29
    Electroluminescent (EL), 29-30
    Gas Plasma, 31
    Liquid Crystal Display (LCD), 30-31

MS-DOS, 53, 59-79
    commands, 63-74
    diagnostics, 77-78
    directory structure, 61-62
    error messages, 74-76
    file management, 76-77
    filenames, 59-60
    memory management, 78-79
    menu building, 79
    virus protection, 78
Operating system, 39, 47
    allocation of resources, 49-51
    classification, 52-53
    function of, 49-50
    interfaces, 52-53
    management of files, 51, 76-77
    monitoring of activities, 51
    number of tasks, 52
    number of users, 52
OS/2, 54-55
Ports, input/output, 26-27
Printer, computer
    daisy wheel, 32-33
    dot matrix, 33-34
    inkjet, 36
    laser, 34-35
Software
    commercial, 38-39
    freeware, 36
    functions of, 39-42
    obtaining, 36-39
    user-supported, 37
Spreadsheets, electronic, 40, 103-119
    advantages, 104
    basic components, 107-109
    commands,
        advanced, 116-118
        standard, 113-116
    definition of, 104
    essential features, 111-113
    hardware requirements, 106-107
    major functions, 109-111
    printing, 110-111, 115-116, 117
    software, 105-106
System (Macintosh), 55-56
Unix, 56
Word processing, 39-40, 81-100
    computer application, 82
    definition of, 82
    hardware requirements, 83-84
    software, 82-83
Word processors,
    features of
        advanced, 92-98
        common, 87-92
    functions of, 84-85
    printing, 85, 91, 98-100